Mechanisms of Environmental Carcinogenesis

Volume II

Multistep Models of Carcinogenesis

Editor

J. Carl Barrett

Head, Environmental Carcinogenesis Group
Laboratory of Pulmonary Pathobiology
National Institute of
Environmental Health Sciences
Research Triangle Park, North Carolina

CRC Press, Inc.
Boca Raton, Florida

Library of Congress Cataloging-in-Publication Data

Mechanisms of environmental carcinogenesis.

    Contents: v. 1. Role of genetic and epigenetic
changes--v. 2. Multistep models of carcinogenesis.
    Includes bibliographies and index.
    1. Carcinogenesis. I. Barrett, J. Carl (James Carl)
[DNLM: 1. Carcinogens, Environmental--adverse effects.
2. Cell Transformation, Neoplastic--chemically induced.
3. Neoplasms, Experimental--chemically induced.
4. Neoplasms, Experimental--familial & genetic.
5. Oncogenes. QZ 202 M48547]
RC268.5.M436   1987   616.99'4071   86-34338
ISBN 0-8493-4670-3 (set)
ISBN 0-8493-4671-1 (v. 1)
ISBN 0-8493-4672-X (v. 2)

    This book represents information obtained from authentic and highly regarded sources. Reprinted material is quoted with permission, and sources are indicated. A wide variety of references are listed. Every reasonable effort has been made to give reliable data and information, but the author and the publisher cannot assume responsibility for the validity of all materials or for the consequences of their use.

    All rights reserved. This book, or any parts thereof, may not be reproduced in any form without written consent from the publisher.

    Direct all inquiries to CRC Press, Inc., 2000 Corporate Blvd., N.W., Boca Raton, Florida, 33431.

© 1987 by CRC Press, Inc.

International Standard Book Number 0-8493-4670-3 (Set)
International Standard Book Number 0-8493-4671-1 (Volume I)
International Standard Book Number 0-8493-4672-X (Volume II)
Library of Congress Card Number 86-34338
Printed in the United States

PREFACE

The process of neoplastic development is now recognized as a multistep process. Specific environmental chemicals can affect this process at unique stages suggesting the involvement of different mechanisms. Recent advances have been made in the understanding of the multistep process of carcinogenesis from experimental studies with animals and cells in culture. In addition, chemicals that affect specific stages of carcinogens have been identified from epidemiological studies of human cancer. The multistep process of carcinogenesis may involve multiple mechanism, epigenetic as well as genetic. The knowledge of the mechanisms of chemically induced genetic and nongenetic processes is to the point that the role of specific mechanisms in carcinogenesis can be discussed. This is exemplified by the study of viral and cellular oncogenes. These volumes present an overview of genetic and epigenetic processes in carcinogenesis. The multistep process of carcinogenesis is described and discussed in chapters on chemical carcinogenesis in animals, epidemiology of human cancers, cell transformation, and viral carcinogenesis. Further discussion of cellular and molecular mechanisms of carcinogenesis is presented in chapters dealing with tumor initiation, tumor promotion, and viral and cellular oncogenes. Examples of specific effects of chemicals and their mechanism of action in the multistep process of carcinogenesis are given and discussed. It is the objective of these volumes to outline possible cellular and molecular mechanisms of carcinogenesis in the context of a multistep process of neoplastic development and to relate these mechanisms to the effects of environmental substances.

THE EDITOR

J. Carl Barrett, Ph.D., is the head of the Environmental Carcinogenesis Group at the National Institute of Environmental Health Sciences, Research Triangle Park, North Carolina.

Dr. Barrett received his Ph.D. degree in Biophysical Chemistry at the Johns Hopkins University in 1974 and then spent 3 years as a postdoctoral fellow in the laboratory of Prof. Paul Ts'o at Johns Hopkins University. During that period he initiated his studies on the role of mutagenesis in carcinogenesis and the multistep process of neoplastic transformation of cells in culture. He has continued in this area and has made significant contributions to the understanding of the mechanisms of environmental carcinogenesis, such as asbestos, diethylstilbestrol, and arsenic, and to the cellular and molecular basis for the different stages on carcinogenesis. His current research efforts are focused on the study of oncogenes and tumor suppressor genes in neoplastic development.

CONTRIBUTORS

Volume I

J. Carl Barrett, Ph.D.
Chief
Environmental Carcinogenesis Section
National Institute of Environmental
  Health Sciences
Research Triangle Park, North
  Carolina

Timothy H. Carter, Ph.D.
Associate Professor
Department of Biological Sciences
St. John's University
Jamaica, New York

Nancy H. Colburn, Ph.D.
Chief, Cell Biology Section
Laboratory of Viral Carcinogenesis
National Cancer Institute
Frederick, Maryland

Peter A. Jones, Ph.D.
Director for Basic Science
Cancer Center
School of Medicine
University of Southern California
Los Angeles, California

Avery A. Sandberg, M.D.
Chief
Department of Genetics and
  Endocrinology
Roswell Park Memorial Institute
Buffalo, New York

Bernard E. Weissman, Ph.D.
Assistant Professor
Hematology/Oncology Division
Childrens Hospital of Los Angeles
Los Angeles, California

Gary M. Williams, M.D.
Associate Director, Naylor Dana
  Institute
Chief
Department of Pathology and
  Toxicology
American Health Foundation
Valhalla, New York

CONTRIBUTORS

Volume II

J. Carl Barrett, Ph.D.
Chief
Environmental Carcinogenesis Section
National Institute of Environmental
  Health Sciences
Research Triangle Park, North
  Carolina

Nicholas E. Day
Unit of Biostatistics
International Agency for Research on
  Cancer
Lyon, Cedex, France

Lennart C. Eriksson, M.D., Ph.D.
Associate Professor
Department of Pathology
Karolinska Institute
Stockholm, Sweden

Emmanuel Farber, M.D. Ph.D.
Professor
Departments of Pathology and
  Biochemistry
University of Toronto
Toronto, Ontario, Canada

William F. Fletcher, B.S.
Graduate Student
Laboratory for Pulmonary
  Pathobiology
Environmental Carcinogenesis Group
National Institute of Environmental
  Health Sciences
Research Triangle Park, North
  Carolina

Henry Hennings, Ph.D.
Research Chemist
Laboratory of Cellular Carcinogenesis
National Institutes of Health
National Cancer Institute
Bethesda, Maryland

John M. Kaldor, Ph.D.
Biostatistician
Unit of Biostatistics and Field Studies
Division of Epidemiology and
  Biostatistics
International Agency for Research on
  Cancer
Lyon, Cedex, France

Joel B. Rotstein, Ph.D.
Research Associate
Department of Pathology and
  Biochemistry
University of Toronto
Toronto, Ontario, Canada

TABLE OF CONTENTS

Volume I

Chapter 1
Genetic and Epigenetic Mechanisms of Carcinogenesis...1
J. Carl Barrett

Chapter 2
Role of DNA Methylation in Regulating Gene Expression,
Differentiation, and Carcinogenesis..17
Peter A. Jones

Chapter 3
Suppression of Tumorigenicity in Mammalian Cell Hybrids.....................31
Bernard E. Weissman

Chapter 4
The Regulation of Gene Expression by Tumor Promoters................................47
Timothy H. Carter

Chapter 5
The Genetics of Tumor Promotion...81
Nancy H. Colburn

Chapter 6
Role of Chromosome Changes in Carcinogenesis..97
Avery A. Sandberg

Chapter 7
DNA Reactive and Epigenetic Carcinogens 113
Gary M. Williams

Chapter 8
Relationship between Mutagenesis and Carcinogenesis................................... 129
J. Carl Barrett

Index.. 143

Volume II

Chapter 9
Cancer Development as a Multistep Process: Experimental
Studies in Animals.. 1
Emmanuel Farber, Joel B. Rotstein, and Lennart C. Eriksson

Chapter 10
Interpretation of Epidemiological Studies on the Context
of the Multistage Model of Carcinogenesis..21
John M. Kaldor and Nicholas E. Day

Chapter 11
Tumor Promotion and Progression in Mouse Skin..59
Henry Hennings

Chapter 12
Cellular and Molecular Mechanisms of Multistep Carcinogenesis
in Cell Culture Models..73
J. Carl Barrett and William F. Fletcher

Chapter 13
A Multistep Model for Neoplastic Development: Role of
Genetic and Epigenetic Changes 117
J. Carl Barrett

Index.. 127

Chapter 9

CANCER DEVELOPMENT AS A MULTISTEP PROCESS: EXPERIMENTAL STUDIES IN ANIMALS

Emmanuel Farber, Joel B. Rotstein, and Lennart C. Eriksson

TABLE OF CONTENTS

I. Introduction 2

II. The Stepwise Development of Hepatocellular Carcinoma in the Rat 2
    A. Models 2
    B. Initiation 4
        1. I-1 — Biochemical Lesion(s) (DNA) 5
        2. I-2 — Fixation by Cell Proliferation 5
    C. Promotion 6
        1. P1 — Microscopic Foci or Islands 6
        2. P2 — Hepatocyte Nodules 7
            a. Biological Properties and Options 7
            b. Biochemical Properties 8
    D. Progression 8
        1. PC-1 — Cell Proliferation Plus Cell Death 8
        2. PC-2 — Cell Proliferation, Cell Death, and Altered Shut-off of Cell Cycle 10
        3. Subsequent Steps 10
    E. Hepatocellular Carcinoma 10
        1. Minimum of Three Steps — HC-1, HC-2, and HC-3 10

III. Mechanisms at Different Steps 11
    A. Initiation 11
    B. Promotion 12
    C. Progression 13

IV. Comparisons with Other Organs and Tissues 14

V. Some Central Questions in Multistep Analysis of Cancer Development 14
    A. Overall Patterns 14
    B. Initiation 14
    C. Promotion 14
    D. Progression 15

VI. Perspectives and Conclusions 15

Acknowledgments 15

References 15

I. INTRODUCTION

We have known for decades that human cancer in the majority of organs and tissues is preceded by reproducible cellular and tissue changes during the 10 to 30 years it usually takes for cancer to develop. In some organs, including the cervix, skin, breast, and bronchial tree, there are tissue and statistical evidence that some of these changes, i.e., atypical hyperplasia, dysplasia, and carcinoma *in situ*, are probably important steps in the carcinogenic process.[1,2]

Since the late 1930s, it has been known that a presumably analogous stepwise process occurs in model experiments on cancer induction with chemicals in the skin of rabbits and mice. These studies led to the paradigm of "two stage carcinogenesis", initiation, and promotion. Since 1971, this was broadened to other organs and tissues beginning with the work of Peraino and colleagues on liver.[3]

Whenever studied, it becomes quickly evident that the designation "two stage" means multistep.[1,2] It is time that the term "two stage" be relegated to historical development, since its use is now both inaccurate and counterproductive in any modern critical and analytical dissection of the carcinogenic process or processes. In this critical review, we shall use the terms "initiation", "promotion", and "progression" for three major overall phases of cancer development[1] and specific designations of individual steps in each phase insofar as these are known or suspected. From the current elementary vantage point or perspective, the carcinogenic process could well involve seven to ten or more steps between an initial cell of origin and the appearance of a malignant metastasizing neoplasm.

Two alternative approaches were seriously considered in the design of this review. The first approach, a brief but systematic analysis of the multistep development of cancer in several organ or tissue systems, was ruled out as too superficial or too general. Such brief or expanded reviews have appeared fairly recently,[1,2] and an alternative approach was selected. One of the more advanced systems the development of hepatocellular carcinoma in the rat, was chosen as a major focus.[4] This system is now yielding to the step-by-step analysis of carcinogenesis and is generating new insights into some key steps in the carcinogenic process. The known steps in the development of this form of liver cell cancer will be delineated briefly. Included will be a short discussion of experimental models and criteria for their utility in any mechanistic analysis. This will be followed by a discussion of known and possible mechanisms for each of the steps. An attempt will be made to correlate the biological with the physiological, the biochemical, and the molecular. Finally, a brief comparison will be made between the liver system and a few other systems (skin, mammary gland, and colon) in an effort to obtain some general principles for the analysis of carcinogenesis in vivo as a multistep process.

The emphasis in this review will be exclusively on chemicals as carcinogens. Although viruses and radiation are known etiological agents in experimental liver cancer and although there is a close association between infection with hepatitis B virus and liver cancer in humans, we have little understanding of the nature of the steps that is seen in these other forms of hepatocarcinogenesis.

II. THE STEPWISE DEVELOPMENT OF HEPATOCELLULAR CARCINOMA IN THE RAT

A. Models

There are at least eight models for the study of liver cancer development with chemicals.[20]

Model 1 — The original model, first developed by Sasaki and Yoshida in 1933 to 1935[5] and studied in great detail by Kinosita[6] and others since, involved observations of liver alterations at many time points during the long-term continuous exposure to a carcinogen in the diet. Azo dyes (o-aminoazotoluene and p-dimethylaminoazobenzene) were most frequently used initially. The lack of synchrony in the development of new cell populations and the inability to dissect the possible relevant changes from the many irrelevant ones make this model virtually unanalyzable for sequence. At best, information about what *might* happen, but not about what *does* happen, could be generated.

Model 2 — A modification of this model, intermittent chronic exposure to a carcinogen, has proven to be very useful in generating large hepatocyte nodules for biochemical and biological studies.[7-9] This model suffers from the same deficiencies as does model 1.

Model 3 — A new type of model, one using an initiator and a promoter, was introduced by Peraino and colleagues in 1971.[3] Phenobarbital has been the most common promoter used, although DDT, PCBs, cyproterone acetate, α-hexachlorocyclohexane, and other inducers of liver enzymes are also effective.[10,11] These chronic enzyme induction models suffer from a common deficiency and lack of synchrony of lesion development, making it difficult to analyze any possible sequence. Foci and nodules appear over many weeks or months, thus preventing the delineation of "what precedes what".

Model 4 — A new concept for a model, the resistant hepatocyte (RH) model, was developed by Solt and Farber in 1976.[12] This was based on the hypothesis (since receiving abundant support) that many chemical carcinogens induce a rare RH during initiation and that such resistant hepatocytes can be rapidly stimulated to develop into focal proliferations (foci) and nodules by brief exposure to low levels of a carcinogen, i.e., 2-acetylaminofluorene (2-AAF), coupled with a mitogenic stimulus. The 2-AAF inhibits proliferation of the vast majority of hepatocytes, the uninitiated (sensitive) ones but not the few resistant ones. In this model, the selection pressure to generate nodules is intense, and the initiated resistant cells thus proliferate synchronously as a cohort. This synchrony lasts for several steps.

Model 5 — Sells and co-workers[13] and Newberne and colleagues[14] have described an additional model in which promotion is effected by the postinitiation exposure to diets deficient in choline and low in methionine or low in all lipotropic components. It appears that the development of nodules in this model is nonsynchronous.

Model 6 — Bannasch and colleagues[15] have described what they call a "stop model" in which rats are exposed to an hepatic carcinogen, i.e., N-nitrosomorpholine, for several weeks and then returned to a basal diet without carcinogen.

Model 7 — Rao et al.[16] and Laurier et al.[17] have described a new model in which orotic acid, the natural precursor in pyrimidine synthesis, is used postinitiation as a promoter. This appears to be a model with good synchrony for foci and nodules.

Model 8 — Poirier and colleagues,[18] Ghoshal and Farber,[19] and Lombardi and associates[87] have found that feeding rats a diet deficient in choline and low methionine without any known addition of a carcinogen induces quite a high incidence of liver cancer.

The following discussion of the known steps involved in the development of liver cell cancer is a composite derived from all of the models. However, because of its synchrony and the brief time periods during which the animals were exposed to the agents, the RH model (model 4) has uncovered the largest number of steps and provided considerable insight into some. The expression of options at certain steps indicates that the permanent "toxic" effects on many cell populations are probably minimal in this model.

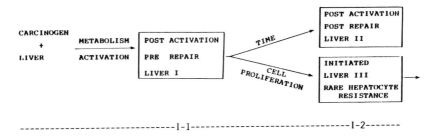

FIGURE 1. Some key steps in initiation of carcinogenesis (I-1 and -2). (Modified from Farber, E., *Cancer Res.*, 44, 5463, 1984. With permission.)

B. Initiation

A major advantage of the rat liver, shared with the mouse skin, for the study of the stepwise development of cancer is the availability of assays for initiation (Figure 1). In 1973, Scherer and colleagues[25] were the first to show that the appearance of islands of hepatocytes with altered histochemical properties could be used as an assay for initiation and that the number of islands was related to the dose of the carcinogen used. This principle has been used to identify islands or foci of altered hepatocytes ("initiated hepatocytes") with many different carcinogens, in several models, and with a variety of histochemical and other markers.[4,27] The early histochemical indexes, especially glucose-6-phosphatase and ATPase, were predominantly negative,[27] i.e., *loss* of a property. Later, positive markers (γ-glutamyltransferase,[32] DT-diaphorase,[33] epoxide hydrolase,[34,35] etc.) increased the versatility of this important approach.

In 1976, a second principle for the identification of initiated hepatocytes was introduced by Solt and Farber[12] and Solt et al.[36] They proposed that one type of initiated hepatocyte was a hepatocyte that acquires, during initiation, a resistance to the inhibitory effects of many carcinogens on cell proliferation (mito-inhibition). They developed an assay for such resistant cells, and this assay has been used to test for potential carcinogenicity in over 60 chemical carcinogens.[12,37,38] The assay selects for carcinogen-induced resistant hepatocytes by rapid growth to form hepatocyte nodules (see below). This assay is the first in any system to use a known physiological change induced by a carcinogen during initiation.

In addition to the assay for resistant hepatocytes, assays for initiation utilize the selection of islands or foci of altered hepatocytes by a promoting environment, i.e., a long-term exposure to phenobarbital, DDT, PCBs or other liver enzyme inducers;[10,11] a diet deficient in choline and low in methionine;[13,14] or orotic acid.[16,17] In these assays, the nature of the property or properties used to select or stimulate the initiated hepatocytes and the basis for the selective effects on these hepatocytes are unknown.

On the basis of our experience in carcinogenesis, we propose the following definition:[39] Initiation is a change in a target tissue or organ, induced by exposure to a carcinogen, that can be promoted or selected to develop focal proliferations, one or more of which can act as sites of origin for the ultimate development of malignant neoplasia.

This definition is singularly devoid of two properties often used in the definition — interactions with DNA and growth. Each is presumptive or invalid. There is an increasing number of chemical carcinogens that do *not* seem to interact with DNA to form adducts. Also, acquisition of any "autonomy of growth" is a much later step and is not seen to accompany initiation in any known organ.

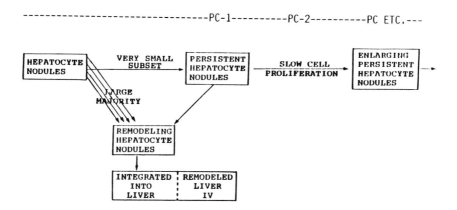

FIGURE 3. Progression of liver carcinogenesis. The early precancerous steps are indicated (PC-1, PC-2, etc.). (Modified from Farber, E., *Cancer Res.*, 44, 5463, 1984. With permission.)

At 2 months postinitiation (PC-1), the percentage of hepatocytes in the persistent nodules undergoing cell proliferation is about 4%, and this increases to 8% by 6 months (PC-2). This is in contrast to the hepatocytes in the liver surrounding the nodules which show a growth fraction of about 0.4%.

At 2 months, not only is there about a tenfold increase in cell proliferation over the surrounding liver, but there is another new phenomenon — cell loss or death. Cell loss or death during the promotion phase of nodule formation (P-2) was virtually nil, with a growth fraction of over 80%.[60] The growth of the nodules matched very well with the rate of cell proliferation and the duration of the cell cycle.[60] In the early persistent nodules (PC-1), it now appears that cell loss or death is a quantitatively significant property of nodules. The 4% growth fraction is almost balanced by a 3% cell loss, thus accounting for a slow rate of enlargement of the nodules.[54,61] This appears to be the first clear-cut appearance of cell death in the sequence in this model and is present throughout the whole nodule-to-cancer sequence as well as in the cancers themselves.[61,62]

Despite the appearance of a spontaneous or autonomous cell proliferation and death, the hepatocytes in the hepatocyte nodules at 2 months show at least some normal control patterns. The cell cycle shows a diurnal rhythm similar to the normal.[61] Also, the nodule hepatocytes respond to the mitogenic stimulus of partial hepatectomy both quantitatively and qualitatively as do normal control or the surrounding liver hepatocytes.[61] The response to partial hepatectomy is vigorous and brisk, and the hepatocytes return to the base line level (4%) by 14 days. The cell cycle phases of the hepatocytes at 2 months, unlike those of the developing nodules during promotion,[60] are normal in that they are identical to those in control untreated animals. At an earlier time, during P-1 and -2, the nodule hepatocytes showed a considerable prolongation of the S phase,[60] but this is no longer present in the 2-month persistent nodule.

At 2 months, the proliferating hepatocytes appear to be a more or less discrete subset of the total nodule hepatocyte population. Instead of showing a random scatter of hepatocytes each undergoing an occasional episode of cell proliferation, the available evidence suggests that the same subset population of nodule hepatocytes undergoes at least two consecutive cell cycles.[61] If future evidence continues to suggest the existence of a proliferating subset, it should become possible to separate such a subset of hepatocytes from the majority of the nodule hepatocytes by flow cytometry. The fact that 60 to 80% of nodule hepatocytes responds to partial hepatectomy by cell proliferation suggests that the majority has retained the quiescence and response pattern of hepatocytes seen in control young adult liver.

2. PC-2 — Cell Proliferation, Cell Death, and Altered Shut-off of the Cell Cycle

At 6 months, the hepatocytes in the nodules have retained their ability to respond briskly and vigorously to partial hepatectomy.[61] Starting from a baseline cell proliferation (growth fraction) of 8%, 60 to 80% of the nodule hepatocytes respond as do the surrounding and control liver hepatocytes to a mitogenic stimulus. However, some nodule hepatocytes have acquired an additional property — lack of shut-off of the cell cycle. Unlike control untreated or surrounding liver or 2-month nodule hepatocytes, a measurable number of 6-month nodule hepatocytes does not return to the baseline level of cell proliferation, but continues to show repeated cell proliferation for at least 3 weeks, in excess of the basal level, in response to partial hepatectomy.[61] Although these findings might be a reflection of a heterogeneous response to a mitogen on the part of the nodule hepatocytes, the available evidence favors the acquisition of a new property by some nodule hepatocytes — a failure to stop cycling once the cell cycle is begun. Again, flow cytometry could prove very useful in obtaining this new subset of nodule hepatocytes.

The persistent nodules at 6 months also show another new property — generation of nodules and hepatocellular carcinoma on transplantation to the spleen.[83] The earlier nodules, like normal liver hepatocytes, grow slowly in the spleen with gradual replacement of the splenic pulp, but without nodules or cancer.

3. Subsequent Steps

Although two new steps relating to the control of cell proliferation have been identified at 2 and 6 months postinitiation, the nature of any subsequent steps remains only speculative and conjectural. Even up to 6 months, the nodule hepatocytes, although showing progressively altered cell cycle control, remain highly predictable and seemingly still under a reasonably rigid control. However, at 2 and 6 months, it is already evident that subsets of nodule hepatocytes are showing new properties that could be interpreted as approaching a more autonomous and more uncontrolled state. Thus, the model is showing what has been suspected for a long time — cellular evolution with small numbers of cells participating or at risk.[63-65] Since the nodule-to-cancer segment of liver cancer development appears to be quite synchronous at least until 6 months, it is anticipated that synchrony might continue, and this would allow a sequential analysis from 6 months to the time of cancer formation.

E. Hepatocellular Carcinoma

1. Minimum of Three Steps — HC-1, HC-2, and HC-3

In most models of hepatocellular carcinoma development with chemicals in adult rats, including the RH model, one usually sees unequivocal malignant neoplasia by 9 to 12 months postinitiation. However, no detailed study of the later steps to cancer has been possible to date. Our experience with the RH model suggests that it may now become feasible, providing the synchrony of lesion development seen at 2, 4, and 6 months is maintained for an additional 2 to 4 months.

It is known that hepatocyte nodules are precursors for hepatocellular carcinoma in at least five studies with different models.[48] This evidence provides clear, convincing evidence that cancer can arise inside hepatocyte nodules. Whether this is an obligatory

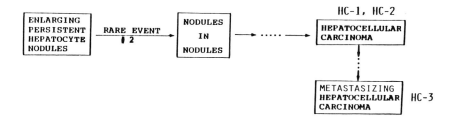

FIGURE 4. Progression of liver carcinogenesis. The later steps, including at least three relating directly to cancer, are indicated. (Modified from Farber, E., *Cancer Res.*, 44, 5463, 1984. With permission.)

requirement for liver cell cancer is, of course, unknown and may well remain so for a long time. However, these observations indicate the utility of studying the nodule-to-cancer sequence as one established sequence (Figure 4).

It is generally seen that the expressions of advanced invasion and metastasis are not always present, by any means, with every unequivocal cancer. In fact, it appears that these are relatively late manifestations of a malignant neoplasm.

However, until the fundamental basis for each property is better understood and sensitive assays are available for the fundamental alterations, it is impossible to discuss intelligently how the properties relate to a stepwise sequence. Until a model is developed for a stepwise analysis, these questions will continue to remain abstract and speculative: Is the order fixed or variable? Is there an interdependence among the three properties in that altered growth must be acquired before the ability to invade and the latter in turn must be acquired before the ability to metastasize?

III. MECHANISMS AT DIFFERENT STEPS

A. Initiation

There is considerable circumstantial but inconclusive evidence that DNA is a major target macromolecule, if not the target. Yet the many dozens or hundreds of molecular changes induced by interactions of a highly reactive ultimate carcinogen with virtually all portions of the cell (including the nucleus, mitochondria, endoplasmic reticulum, membranes, etc. and DNA, RNAs, and proteins) make the scientific evidence for the exclusive role of any single target in carcinogenesis unobtainable so far. Even if it is assumed that DNA is the major or only target, the number of chemical, functional, and steric alterations in the DNA of target tissues[21,22] again offers a virtually insoluble problem if a relationship between any single one or even a single group of alterations and initiation of carcinogenesis is sought. The number of parallel phenomena that accompany the critical one or ones is no doubt large.[1,86]

Therefore, any conclusion concerning the role of mutation, gene rearrangement, amplification, suppression, activation, and oncogene activity, etc. must remain at best highly speculative.

Biologically, in vivo initiation is associated with the appearance, in a rare cell, of a change that can be made permanent or relatively so by a round of cell replication.[68] These properties per se are consistent with a mutation-like change in DNA that can be fixed by one round of DNA replication. However, other types of DNA-associated phenomena, i.e., gene rearrangements or activations, might also fit the biological "facts" as presently known. Whatever the nature of the relevant molecular lesion, it is *not* related directly to any autonomous or spontaneous growth.

In the liver, one or more of these presumed changes in DNA must generate a new biochemical pattern, which is associated with resistance to xenobiotics, the potential for a new state of differentiation, and a genetic program for redifferentiation. Clearly, this can only reside in some regulatory gene of a major type. This regulator gene could be structurally altered by a mutation or functionally altered by a change in some related region of the DNA. By allowing for a biochemical program for resistance to the cytotoxicity of xenobiotics including carcinogens, this may be sufficient to explain the first major phase of carcinogenesis, the genesis of focal proliferations or hepatocyte nodules.

Thus, by an intensive search for the relevant genes for one or more phenotypic components of hepatocyte nodules, it becomes theoretically possible to begin to focus on the key gene or genes, the alteration of which relates to initiation. The 21-kdalton polypeptide in nodule cytosol[57] could be one such penetrating probe. From the current perspective, it appears much more realistic and profitable to work backwards[1] from the known to the unknown if we are to understand the essential nature of the key interactions of a chemical carcinogen with DNA that relates to the first phases of cancer development. The forward, head-on approach by looking for changes in DNA without a guide would be an excellent example of seeking the needle in the haystack.

B. Promotion

The liver offers one established mechanism for promotion, differential inhibition.[27] By providing a stimulus for cell proliferation (partial hepatectomy or cell death by CCl_4) and by inhibiting cell proliferation of the vast majority of uninitiated susceptible hepatocytes, the few rare resistant hepatocytes (the initiated ones) are able to respond to the mitogenic stimulus and rapidly generate focal proliferations or nodules. Thus, in this system, promotion is essentially by *inhibition*, not by selective *stimulation*. In humans, since cancer often arises in an atrophic, not a hyperplastic organ or tissue (e.g., stomach, liver, or pancreas), it is likely that this overall mechanism may well be a common one. This is certainly most likely in animals and humans where cancer arises in response to repeated long-term exposure to one or more carcinogens, i.e., the carcinogen is both initiator and promoter. This pattern appears to be much more common in nature than is the initiation-promoter pattern that is being used more and more experimentally.

Other patterns of promotion can be proposed;[27] for example, differential stimulation or differential recovery are two feasible possibilities.

It should be emphasized that the current research approaches for the study of promoters, like that of initiators, are the forward one. By cataloging what biochemical, molecular, genetic, immunologic, or biophysical changes are induced by a particular agent, the mechanism of promotion is hopefully understood. To date, this approach has generated virtually no insight into the mechanisms of promotions, and the prospects that it may do so soon seem remote. However, this approach has reinforced in a clear manner the existence of synergisms among promoters and the dissection of promotion in mouse skin into at least two steps.[71]

The likelihood of understanding a mechanism before a phenomena is delineated seems truly unrealistic. For example, since differential inhibition is one mechanism for promotion in liver carcinogenesis, it would be irrational to focus on the presence of a selective proliferative response by the initiated hepatocytes and ignore the failure to respond by the majority. A more rational approach, which is productive, is to study the biochemical and genetic basis for the resistance by the initiated hepatocytes.

From the RH model, it is also apparent that populations derived from the same cells, presumably by clonal expansion, show different properties with different degrees of expansion. Initiated RH are stable for months as small invisible or visible cells or foci.[68] However, on expansion to nodules, the resistant initiated hepatocytes demonstrate a new biological option — remodeling by redifferentiation.[55] Thus, two clones of the same cells, expanded to different degrees, show two different patterns of behavior.

Also, it is well-known in embryology or developmental biology that clonal expansion plays an important role in differentiation.[69] In muscle and other types of cells, the concept of quantal cell cycle is very attractive. Cells seem to require a certain number of cell divisions before a new pattern of gene expression can become manifest. This could well be the pattern in the carcinogenic process where different degrees of clonal expansion of a similar cell population generate new focal proliferations of different biological behavior. If this is applicable to the process of cancer development,[69] it is not surprising that smaller foci or islands of initiated cells may have quite different biological options available than do larger nodules. This principle should be explored further both in vivo and in vitro.

C. Progression

The RH model has brought into focus two aspects of the possible mechanisms of progression to cancer: (1) cell death as a trigger or consequence of cell proliferation and (2) the production of growth factors and stimulation of oncogene expression.

The sudden appearance of cell death at the earliest observable time in the persistent nodule, i.e., in the nodule to cancer sequence, introduces an important component in carcinogenesis in vivo. Can cancer development occur without cell death? Does the cell death trigger the spontaneous or autonomous cell proliferation in step PC-1, or is cell death a consequence of the cell proliferation? The new availability of a model to study this might lead to some answers.

Another important consideration linked to cell death as a trigger or consequence of cell proliferation is whether step PC-1 is associated with the new production of growth factors, perhaps on an autocrine basis.[70] Since some oncogenes are related to some growth factors,[70] it is possible that one or more oncogenes may become expressed or show increased expression at step PC-1. Again, the RH model makes it now possible to pose this question.

Regardless of the basis for the spontaneous cell proliferation in steps PC-1, -2, and later, the continual proliferation of a small population in the persistent nodules without constant shedding of the progeny (such as occurs on a surface or in bone marrow) may now make these steps open to a process of cell evolution[63] by mutation and selection[64] or clonal evolution,[65] and these may well become the major overall mechanisms for the progressive pressure toward more and more malignant behavior.

An important aspect of this phase of carcinogenesis is the study of increasing growth, invasive properties, and metastasis as separate sets of mechanisms. It is well known that these three sets are acquired separately and seriatim. Without a knowledge of these three apparently separable phenotypic expressions, any hope of studying the basis for the great diversity and heterogeneity of malignant neoplasms seems remote. So far virtually all of the studies on malignant neoplastic populations, be they monoclonal or polyclonal, have generated little new insight into the essences of malignancy with one notable exception — the possible genesis of growth factors (so-called transforming growth factors, etc.) and the interesting possibility of autocrine control of cell proliferation by malignant neoplastic cells.[70] The vast majority of phenotypic changes seen in cancer might well be epiphenomena, "noise", a fundamental base of growth, invasion, and metastasis. Studies of cell populations in which these are not delineated and controlled as variables are not likely to be fruitful of new insights.

IV. COMPARISONS WITH OTHER ORGANS AND TISSUES

The pattern of multistep carcinogenesis in the rat liver is remarkably similar in principle to multistep development of cancer in mouse skin,[71] urinary bladder,[72] in the colon in humans and mice,[73,74] in the pancreas,[75,77,81] mammary gland,[78] and respiratory tract.[79] The interactions with DNA and other cell constituents, the importance of cell proliferation, and the clonal expansion during the promotion phase appear to be very common. The reversibility of many focal proliferations (papillomas, nodules, and polyps) appears to be a common phenomenon in some systems.[1,2] Also, the multistep nature (at least two steps) of the promotion phase in the skin is well-documented.[71] The stepwise nature of progression has been poorly studied and requires increased emphasis if we are ever to understand the steps between focal proliferations and malignant neoplasia.

The genesis of melanoma in the human[80] also has many resemblances to the basic patterns in the rat liver. In fact, some of the steps in the human melanocyte and rat hepatocyte systems seem almost superimposable. These similarities are encouraging in that they support the thesis that the detailed elucidation of *any* system, be it human or animal, will almost certainly act as a blueprint and a model, in principle, for many other systems, both clinical and experimental.

V. SOME CENTRAL QUESTIONS IN MULTISTEP ANALYSIS OF CANCER DEVELOPMENT

Given the relatively undeveloped understanding of the multistep nature of cancer development in all systems, it becomes important to concentrate on, or at least to formulate, some of the key questions that must be studied for clarification of the most fundamental aspect of cancer research, the stepwise development of cancer.

A. Overall Patterns

The studies in experimental animals are focused almost exclusively on one pattern of cancer development, the multistep pattern with focal proliferations.[20] There are at least three patterns,[20] only one of which has papillomas, polyps, etc. as key precursors. Are there truly systems of cancer development without such focal proliferations as cancer precursors?

B. Initiation

To date, initiation has been demonstrated in many tissues or organs as well as in vitro. However, a central issue concerns the relation of the initiated events to cancer. Does an initiating exposure to a carcinogen convey any message that pertains to the ultimate cancer, or does it only induce the first step in the few initiated cells, i.e., the potential to become the site for a focal proliferation? The simplest hypothesis is the latter — an initiator inducing a change allowing that cell to become clonally expanded by a further exposure to the same, or a different carcinogen or promoter. Intuitively, most cancer researchers assume that an initiating carcinogen not only initiates but induces some effect that relates more directly to cancer.

C. Promotion

In most models of carcinogenesis, focal or clonal expansion of the few initiated target cells is the only immediate visible concomitant of promotion. If the expansion is sufficient in degree, is this all that is associated with promotion, or must there be some additional alteration in the expanding cell population that relates to cancer? Additional so-called hits are commonly discussed in this context, without any experimental evidence.[26,82] The possible nature of the so-called multihits have not been discussed, other than as additional mutations.

D. Progression

What signals the subset of persistent nodule hepatocytes or similar subsets in other systems to begin to undergo cell proliferation? The fundamental issue to be asked is this secondary to cell death as a new acquired property at the PC-1 step or are the new cells now generating growth factors, perhaps because of the activation of some selected oncogenes? The steps PC-1 and -2 are the places to ask these questions, since the system is uncomplicated by uncontrolled growth, invasion, and metastasis, and by the many epiphenomena that in concert contribute to the bewildering array of phenotypic expressions in any fully developed cancer.

VI. PERSPECTIVES AND CONCLUSIONS

Clearly, the stepwise analysis of a carcinogenic process opens up many new possibilities for study and generates crucial new questions relating to our understanding of the way in which cancer develops mechanistically.

The remarkable similarities between many models in different organs and tissues, with different carcinogens, and in humans, mice, rats, and other animals, coupled with the paralyzing diversity and heterogeneity of any end-stage cancer, point clearly to the importance of studies in cancer development as basic to our ultimate understanding of cancer. This obvious conclusion seems not to have been taken to heart by the vast majority of cancer researchers, given the relative paucity of studies in the stepwise analysis of cancer development. The experience with one system, the liver, indicates that new analyzable features of neoplasia will continue to be generated by expanded studies on the steps through which cells evolve as they slowly move toward cancer.

ACKNOWLEDGMENTS

The research of the authors included in this chapter was supported by grants from the Public Health Service CA-21157, National Cancer Institute, National Cancer Institute of Canada, Medical Research Council of Canada (MT-5994), and the Swedish Medical Research Council. We would like to express our sincere thanks to Lori Freund for her help in preparation of the manuscript.

REFERENCES

1. Foulds, L., *Neoplastic Development,* Vol. 1 and 2, Academic Press, New York, 1969, 1975.
2. Farber, E. and Cameron, R., The sequential analysis of cancer development, *Adv. Cancer Res.,* 35, 125, 1980.
3. Peraino, C., Fry, R. J. M., and Staffeldt, E., Reduction and enhancement by phenobarbital of hepatocarcinogenesis induced in the rat by 2-acetylaminofluorene, *Cancer Res.,* 31, 1506, 1971.
4. Emmelot, P., Developmental phases in liver carcinogenesis, *Biochim. Biophys. Acta,* 605, 149, 1980.
5. Sasaki, T. and Yoshida, T., Experimentelle Erzeugung des Lebercarcinoms durch Futterung mit O-Amidoazotoluol, *Virchow's Arch. Pathol. Anat. Physiol. Klin. Med.,* 295, 175, 1935.

6. Kinosita, R., Studies on the carcinogenic chemical substances, *Trans. Jpn. Pathol. Soc.*, 27, 665, 1937.
7. Reuber, M. D., Development of preneoplastic and neoplastic lesions of the liver in male rats given 0.025 percent N-2-fluorenylacetamide, *J. Natl. Cancer Inst.*, 34, 697, 1965.
8. Epstein, S., Ito, N., Merkow, L., and Farber, E., Cellular analysis of liver carcinogenesis: the induction of large hyperplastic nodules in the liver with N-2-fluorenyacetamide or ethionine and some aspects of their morphology and glycogen metabolism, *Cancer Res.*, 27, 1702, 1967.
9. Teebor, G. W. and Becker, F. F., Regression and persistence of hyperplastic nodules induced by N-2-fluorenylacetamide and their relationship to hepatocarcinogenesis, *Cancer Res.*, 31, 1, 1971.
10. Pitot, H. C. and Sirica, A. E., The stages of initiation and promotion in hepatocarcinogenesis, *Biochim. Biophys. Acta,* 605, 149, 1980.
11. Schulte-Hermann, R., Ohde, G., Schuppler, J., and Timmermann-Trosiener, I., Enhanced proliferation of putative preneoplastic cells in rat liver following treatment with the tumor promoters phenobarbital, hexachlorocyclohexane, steroid compounds and nafenopin, *Cancer Res.*, 41, 1556, 1981.
12. Solt, D. B. and Farber, E., New principle for the analysis of chemical carcinogenesis, *Nature (London)*, 263, 702, 1976.
13. Sells, M. A., Katyal, S. L., Sell, S., Shinozuka, H., and Lombardi, B., Induction of foci of altered γ-glutamyl-transpeptidase positive hepatocytes in carcinogen-treated rats fed a choline-deficient diet, *Br. J. Cancer*, 40, 274, 1979.
14. Newberne, P. M., Rogers, A. E., and Nauss, K. M., Choline, methionine and related factors in oncogenesis, in *Nutritional Factors in Oncogenesis and Maintenance of Malignancy,* Butterworth, P. E. and Hutchinson, M. L., Eds., Academic Press, New York, 1983, 247.
15. Bannasch, P., Moore, M. A., Klimek, F., and Zerban, H., Biological markers of preneoplastic foci and neoplastic nodules in rodent liver, *Toxicol. Pathol.,* 10, 19, 1982.
16. Rao, P. M., Nagamine, K., Ho, R. K., Roomi, M. W., Laurier, C., Rajalakshmi, S., and Sarma, D. S. R., Dietary orotic acid enhances the incidence of γ-glutamyltransferase positive foci in rat liver induced by chemical carcinogens, *Carcinogenesis*, 4, 1541, 1983.
17. Laurier, C., Tatematsu, M., Rao, P. M., Rajalakshmi, S., and Sarma, D. S. R., Promotion of orotic acid of liver carcinogenesis in rats initiated by 1,2-dimethylhydrazine, *Cancer Res.*, 44, 2186, 1984.
18. Mikol, Y. B., Hoover, K. L., Creasia, D., and Poirier, L. A., Hepatocarcinogenesis in rats fed methyl-deficient amino acid defined diets, *Carcinogenesis,* 4, 1619, 1983.
19. Ghoshal, A. K. and Farber, E., The induction of liver cancer by dietary deficiency of choline and methionine without added carcinogens, *Carcinogenesis*, 5, 1367, 1984.
20. Farber, E., Pre-cancerous steps in carcinogenesis: their physiological adaptive nature, *Biochim. Biophys. Acta,* 738, 171, 1984.
21. Glover, P. L., Ed., *Chemical Carcinogens and DNA*, Vol. 1 and 2, CRC Press, Boca Raton, Fla., 1979.
22. Rajalakshmi, S., Rao, P. M., and Sarma, D. S. R., Chemical carcinogenesis: interactions of carcinogens with nucleic acids, in *Cancer: A Comprehensive Treatise*, Vol. 1, 2nd ed., Becker, F. F., Ed., Plenum Press, New York, 1982, 335.
23. Farber, E., Perspectives in cancer research. The multistep nature of cancer development, *Cancer Res.*, 44, 4171, 1984.
24. Cayama, E., Tsuda, H., Sarma, D. S. R., and Farber, E., Initiation of chemical carcinogenesis requires cell proliferation, *Nature (London)*, 275, 60, 1978.
25. Scherer, E., Hoffmann, M., Emmelot, P., and Friedrich-Freksa, H., Quantitative study of foci of altered liver cells induced in the rat by a single dose of diethylnitrosamine and partial hepatectomy, *J. Natl. Cancer Inst.*, 49, 93, 1972.
26. Emmelot, P. and Scherer, E., The first relevant cell stage in rat liver carcinogenesis: a quantitative approach, *Biochim. Biophys. Acta,* 605, 247, 1980.
27. Farber, E., The sequential analysis of liver cancer induction, *Biochim. Biophys. Acta,* 605, 149, 1980.
28. Columbano, A., Rajalakshmi, S., and Sarma, D. S. R., Requirement of cell proliferation for the initiation of liver carcinogenesis as assayed by three different procedures, *Cancer Res.*, 41, 2079, 1981.
29. Ying, T. S., Sarma, D. S. R., and Farber, E., Role of acute hepatic necrosis in the induction of early steps in liver carcinogenesis by diethylnitrosamine, *Cancer Res.*, 41, 2096, 1981.
30. Craddock, V. M., Cell proliferation and experimental liver cancer, in *Liver Cell Cancer,* Cameron, H. M., Linsell, D. A., and Warwick, G. P., Eds., Elsevier, Amsterdam, 1976, 152.
31. Farber, E., The pathology of experimental liver cancer, in *Liver Cell Cancer,* Cameron, H. M., Linsell, D. A., and Warwick, G. P., Eds., Elsevier, Amsterdam, 1976, 243.
32. Kalengayi, M. M. R., Ronchi, G., and Desmet, V. J., Histochemistry of gamma-glutamyl transpeptidase in rat liver during aflatoxin B_1-induced carcinogenesis, *J. Natl. Cancer Inst.*, 55, 579, 1975.

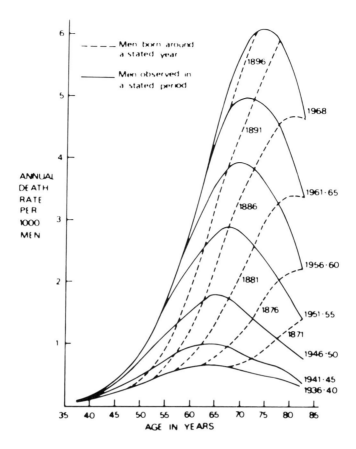

FIGURE 2. Age-specific male lung cancer mortality in U.K., showing the curves obtained by calendar period (solid line) and birth cohort (dashed line). (From Doll, R., *J. R. Stat. Soc. A*, 134, 133, 1971. With permission.)

present and additive on a logarithmic scale, they cannot be distinguished from each other or the age effect in a single population.[40]

Another factor which could distort age-incidence curves is variation in susceptibility due to host factor differences. For example, melanoma incidence is known to vary widely by skin type, even among Caucasian populations, with those with fair skin and light hair being at highest risk.[41-43] At the cellular level, this is presumably related to differences in the response of the melanocyte to UV irradiation or the amount of irradiation received by the melanocyte, and may be mathematically expressed as a variation among individuals in the transition rates between stages (λ_i in Equation 1; μ_i in Equation 2). An age-incidence curve obtained from a heterogeneous population will resemble that of the higher risk segment initially and converge to that of the lower risk segment. Variations within a population in the exposure level to environmental carcinogens can obviously have a similar effect.

B. Interpretation of Estimated Age-Incidence Curves under the Multistage Model

Having stated some of the important difficulties associated with the estimation of the age-incidence relationship, it is nonetheless worth describing the work which has been done on age-incidence curves and the extent to which they agree with the predictions of the multistage model.

The most exhaustive study of the effect of age on cancer incidence is still that of Cook et al.,[13] who took data from 11 cancer registries on 24 cancer sites (7 in both sexes) reported in *Cancer Incidence in Five Continents*.[33] Using maximum likelihood estimation, Equation 1 was fitted for each combination of site and registry, and the goodness-of-fit of the relationship tested. Only the age group 35 to 75 was considered, and the sites chosen were those for which incidence was known to increase in incidence with age throughout adult life. The authors concluded that "the simple power relationship between cancer incidence and age is inadequate to explain the greater part of the observed data," since for about one third of the approximately 340 curves examined, there was significant departure from linearity on a log-log scale. However, the authors felt that apart from those for cancer of the larynx, corpus uteri, and ovary, the plots were sufficiently close to linearity to permit the estimation of the slope k, one less than the putative number of stages. The highest average value (about 11) was for prostatic cancer, and the lowest was for melanoma (about one), but most cancers fell in the range of 4 to 6. It was found that the value of k was related much more strongly to the type of cancer than to the sex or population from which the rates were drawn. This fact would seem to suggest that the index k is summarizing something of biological importance which is common to specific cancers, if not necessarily the number of stages in a multistep carcinogenic process. However, it should be emphasized that only seven cancers were considered in both sexes.

Cook and co-authors, in addition to expressing far more caution about the generality of the log-log relationship than many of the subsequent authors who have quoted their paper, provide a deep and thoughtful discussion of the possible modifying factors unaccounted for in their analyses. The most important of these is almost certainly the effect of cohort and/or period on the age-incidence curves.

Various means have been used by other authors to estimate the age-incidence curve in a manner which is unaffected by factors of this kind. Doll[12] constructed the age-incidence curve for lung cancer in lifelong nonsmokers, combining data from one British[44] and two American studies.[45,46] Since cigarette smoking is believed to be the overwhelming risk factor for lung cancer in the U.K. and the U.S., the figures for nonsmokers could be expected to be free of important effects other than those related to age. Figure 3 shows the resulting curve on a double log plot, and indeed it is strikingly linear, with a slope of about four. The figure also shows age-incidence curves for smokers, which are discussed in more detail in Section IV.A.1.

If one is prepared to consider more than one population or more than one cancer site simultaneously and make certain assumptions about the age, period, or cohort effects, it is possible to effectively adjust age effects for those of cohort and period. Day and Charnay[47] suggested that the effect of age on incidence could be assumed to be constant (multiplicatively) across genetically similar populations, thereby ascribing differences in incidence among the populations to cohort and period effects. Using this method, the lung cancer age-incidence curve was estimated using data from Finland and Slovenia, resulting again in a quite convincing linear log-log relationship (the slope was not estimated). In studying melanoma incidence in Norway, Boyle et al.[39] considered models where either the age or the cohort effects were assumed to be constant across subsites of the body. In the model with constant cohort effects, the estimated age-incidence curves were roughly linear on the log-log scale with a slope of about four, but tended to flatten in older age groups for sites other than the head and neck.

The two-stage model proposed by Moolgavkar and Knudson[21] has been applied to a number of cancers which were clearly not of the log-log kind. By incorporating a priori information (some known and some speculative) on breast tissue growth kinetics, Moolgavkar et al.[48] fitted the two-stage model rather well to breast cancer incidence

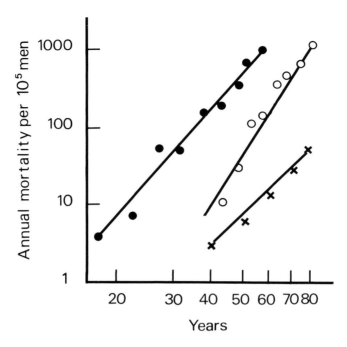

FIGURE 3. Age-specific lung cancer mortality for nonsmokers and smokers. (●—● = cigarette smokers by duration of smoking; ○—○ = cigarette smokers by age; x—x = nonsmokers by age.) (From Doll, R., *J. R. Stat. Soc. A*, 134, 133, 1971. With permission.)

data from six different registries. Because of major secular increases in incidence in some populations, it was first necessary to standardize the age-incidence curves for cohort effects.[38] Figure 4 shows the fit obtained for data on breast cancer obtained from six registries which exhibited quite different cross-sectional age-incidence curves before this standardization was applied. The model suggested by Pike[32] provides an equally good fit to the observed age-incidence curve of breast cancer among white females in the U.S.

One group of cancers whose incidence rates are clearly not linearly related to age on a double logarithmic scale is those which seems to be associated with viral agents. Figure 5 shows the age-specific rates of nasopharyngeal carcinoma among Singapore Chinese,[49] cervical cancer rates in Cali, Colombia and Bombay, India,[36] and Burkitt's lymphoma.[50] Hepatocellular carcinoma appears to increase in a log-log fashion with age in some registries from regions where the associated hepatitis B virus is endemic, while other registries report a downturn similar to that observed for nasopharyngeal cancer.

IV. THE MULTISTAGE MODEL AND EXPOSURES OF DEFINED DURATION AND DOSE

A. Predictions of the Multistage Model for Carcinogenic Exposures of Defined Duration and Dose

It soon became clear to epidemiologists that only a limited amount of insight into the mechanisms of carcinogenesis could be obtained from studying population age-incidence curves in isolation. They therefore looked around for other types of data to validate their models and, in particular, focused upon situations where humans had

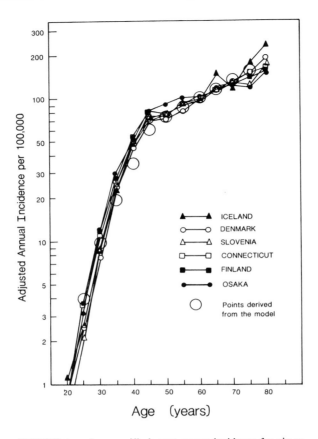

FIGURE 4. Age-specific breast cancer incidence for six registries and the fit obtained using the two-stage model. (From Moolgavkar, S. H. and Knudson, A. G., Jr., *J. Natl. Cancer Inst.*, 66, 1037, 1981. With permission.)

been exposed to known carcinogens over defined time periods. Cigarette smoking is probably the best quantified of known carcinogenic exposures, in addition to being by far the most important etiologic agent yet identified for human cancers. The large follow-up studies on British doctors,[14,44,51] American veterans,[45] and a large group of American volunteers[46] have provided detailed information on cancer mortality as it relates to the age, level, duration, and type of cigarette consumption. Another carcinogenic exposure which has been intensively studied in a heterogeneous population is radiation following the atomic bombing of Hiroshima and Nagasaki.[52] Other quantifiable exposures have occurred in much more specialized groups of individuals, i.e., industrial workers or patients undergoing treatments for specific diseases.

In this section, we examine the predictions of the multistage model under such well-defined exposure situations. These predictions will be compared with epidemiological observation in Section IV.C, following a discussion in Section IV.B of the difficulties involved. We consider a carcinogenic exposure which commences at age t_0 and stops at age t_1. A number of authors[15,16] have derived mathematical expressions for the cancer incidence which would be predicted by the Armitage-Doll multistage model following t_0. In the general case, when multiple stages may be affected by the carcinogen and the level of exposure is variable, these expresssions are rather complex and difficult to apply to the interpretation of epidemiological data. Most authors have therefore con-

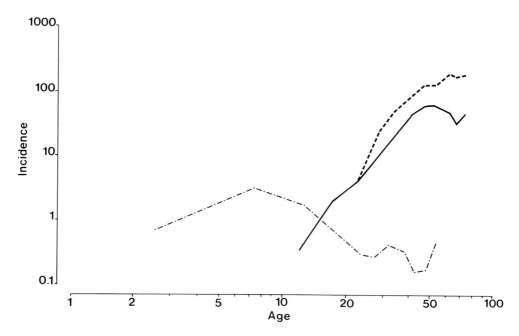

FIGURE 5. Annual age-specific incidence per 100,000 of nasopharyngeal carcinoma among Singapore males (—), cervical cancer in Cali, Colombia (- - -), and Burkitt's lymphoma among males in Uganda (-·-·-).

sidered the idealized situation in which the carcinogen only increases the transition rate for one stage. It is further assumed that the increase occurs instantaneously to a new, constant level during exposure and that when exposure ceases, the transition rate for the affected stage returns to the constant preexposure level. When the transition rate affected is the first or the penultimate, it is possible to obtain very simple formulas for the excess cancer incidence resulting from exposure patterns of this kind. If some other transition is affected, the corresponding formulas are much more cumbersome. Therefore, since we have no idea which transition rates are increased by the exposure, agents affecting the first or penultimate stages have been considered as representations of the more general ideas of "early"- and "late"-stage carcinogens, respectively. These concepts have in turn been viewed as initiation and promotion, but it is not clear whether these concepts have a direct analogue under the Armitage-Doll model.

The excess cancer incidence predicted by the model for an early- or late-stage carcinogen differs in a number of fundamental aspects. First of all, the increase in cancer incidence due to an agent affecting late transitions would be predicted to occur much earlier than the increase following an early-stage carcinogen, since in the former case the extra cells transformed by the exposure are temporally closer to the final, fully malignant stage. If an agent increases the rate of the final transition, the incidence should increase immediately following exposure, in contrast to a first-stage carcinogen whose effect would not be expected to be manifested for many years after first exposure. Figure 6 plots theoretical curves for the cancer incidence following exposure to agents which affect only one of the stages in a five-stage process which has equal transition rates in the absence of exposure. It is assumed that exposure begins at age 20, is continuous thereafter, and that the exposure increases the single transition rate by a factor of ten. The increase in incidence for the early-stage agent is not apparent until about 10 years after exposure began, and even for an agent affecting the second

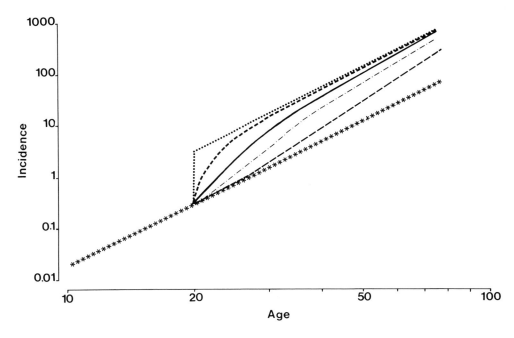

FIGURE 6. Predicted age-specific cancer incidence following continuous exposure beginning at age 20 to a constant level of agents which act on only one of the stages of a five-stage process. The number indicates the stage affected. 1 = — — —, 3 = ———, 5 = · · · · .

transition it takes 10 years before the incidence is increased by twice the background. The risk continues to rise at a faster rate than the background, so that the relative risk increases continuously during exposure. In contrast, the incidence following exposure to a penultimate-stage agent rises rapidly and reaches an asymptote parallel to the incidence curve for unexposed individuals after about 20 years. On the log-log scale, this results in two parallel lines and may also be expressed as a constant relative risk. For the last-stage agent, incidence is instantaneously increased to the upper line. It is important to emphasize that for both early- and late-stage carcinogens, the excess risk increases with duration of exposure.

There is a corresponding contrast in the pattern of excess incidence predicted for the two types of agents when exposure ceases. For a carcinogen whose effect has been to increase the rate of early-stage transitions, the incidence of cancer following cessation of exposure will continue in excess for a considerable length of time almost as if exposure had not stopped; although its rate of increase will gradually slow, the excess cancer incidence will rise indefinitely. However, when exposure to a last-stage carcinogen is terminated, the excess incidence immediately returns to zero because there is no longer any additional pressure on penultimate-stage cells to make the final transition. When exposure to an agent affecting penultimate-stage transitions is removed, the excess incidence should remain fixed at the level attained when exposure stopped, since the result of exposure is an increase in the number of cells ready to make the final transition to malignancy. Figure 7 shows log-log plots of cancer incidence following 20 years of exposure beginning at age 20 to five agents, each affecting only one of the stages in a five-stage process in the same way as in the previous figure. Again, it should be mentioned that the excess risk increases with duration of exposure for both types of agents, even if exposure has ceased.

Another variable which has been used to distinguish agents acting at different stages

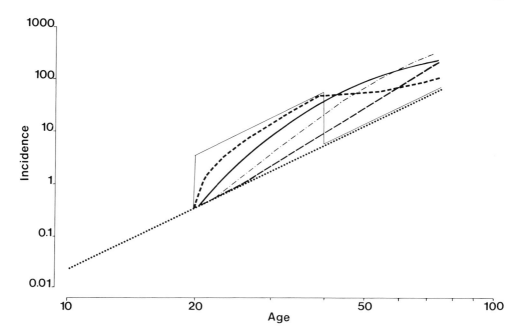

FIGURE 7. Predicted age-specific cancer incidence after cessation of exposure at age 40 years to agents which act on only one of the stages of a five-stage process. The number indicates the stage affected, with zero being the background. 0 = ****, 1 = — — —, 2 = ·-·-·-·, 3 = ———, 4 = - - - -, 5 = ····.

of the carcinogenesis process is the effect due to age at which exposure occurred. Under the Armitage-Doll model, it is assumed that in a particular tissue the number of normal cells which are susceptible to transformation is effectively constant, at least through adult life. Therefore, an exposure which increases transition rates to the first stage will have the same effect on cancer incidence, regardless of when in life exposure occurs. On the other hand, the number of first- and later-stage transformed cells is assumed to be continually increasing, due to "background" transition processes taking place in the tissue. Thus any agent which acts on these cells to increase the rate of transitions to the next stage would be expected to have a greater effect if applied later in life when there are more such cells. Figure 8 shows the excess cancer incidence following continuous exposure to agents affecting only the first, third, or last of the five stages, when the exposure begins at age 20, 25, and 30, respectively. Note that the x-axis indicates time since exposure began, rather than age, so that the curves for exposure beginning at different ages have been superimposed.

In addition to these temporal variables, the other important determinant of risk is the level of exposure whose relationship to risk can also be interpreted in terms of the multistage model. However, unlike the time variables which are reasonably easy to obtain accurately in epidemiological studies, reliable measurements of exposure are almost nonexistent. Even for the few special cases for which usable information on exposure levels is available (e.g., tobacco smoking, asbestos, ionizing radiation, alcohol consumption, and certain therapeutic drugs), there are major deficiencies in the data. For example, while it may be possible to record with reasonable accuracy the number of cigarettes smoked per day, such dose-related factors as the amount of inhalation, fraction of each cigarette smoked, and changes in the brand or composition of the cigarettes are very difficult to quantify. Nevertheless, it is possible to derive theoretical results regarding the dose-response relationships which would be expected for carcin-

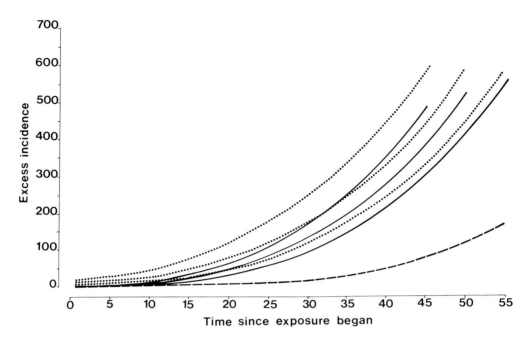

FIGURE 8. Predicted excess cancer incidence following continuous exposure commencing at ages 20, 25, and 30 to agents which affect only the first, third, or fifth stage of a five-stage process. The number indicates the stage affected, with zero being the background. 0 = ****, 1 = — — —, 2 = ··-··-··, 3 = —, 4 = - - - -, 5 = ····.

ogens acting at different stages. It is reasonable to suppose that the effect of a low dose on a single transition rate would be linear, since by definition a transition between stages should be a one-hit process, whether occurring "spontaneously" or due to an external agent reacting with the cellular DNA. At higher doses, we might expect the influence of cell killing, the induction of repair enzymes, or other factors to modify the relationship. Under the linear assumption, the relationship between dose and cancer incidence following exposure to an agent which only affected transition to one stage would also be linear, no matter which stage was affected. However, if an agent affected more than one stage, higher powered dose-incidence relationships would be expected, with the power determined by the number of stages affected. An important feature of the multistage formulation is that the transition rates are related to the instantaneous dose rate, rather than any cumulative measure of dose. Thus, cumulative dose measures such as pack/years for cigarette smoking have little meaning in the context of the model. Duration of exposure is of course an important variable, but generally cannot be viewed symetrically with dose rate.

Further complexity is added by considering the effect on cancer incidence of exposure to two different carcinogens. Siemiatycki and Thomas[53] demonstrate by a series of examples that even in the hypothetical case where exposure to two agents occurs simultaneously and continuously, a wide variety of interactive effects could be predicted, depending on the stage or stages affected by each agent and possible biochemical reactions between the agents themselves. Two agents which do not interact biochemically with each other would be predicted to have an additive effect on excess cancer risk if they acted on cells in the same stage to increase the transition rate, and a multiplicative effect if they acted on cells in different stages. If one agent acts on only one stage and the other acts on both the same stage and another stage, the effect on excess cancer incidence would be between multiplicative and additive.

Under the two-stage model proposed by Moolgavkar and Knudson,[21] a distinction is made between initiators and promoters on the basis of their effect on the parameters in Equation 2. Agents which increase the transition rates μ_1 or μ_2 are initiators, while promoting activity is ascribed to agents which favor proliferation of intermediate cells, in that they increase the net growth rate expressed by the difference $\alpha - \beta$. Cardis[55] has made a detailed study of the cancer incidence predicted by this model under various types of exposures.

B. Statistical Issues in the Interpretation of Epidemiological Observations

Although one can predict the pattern of cancer incidence using the multistage model, there still remain fundamental problems in comparing the predictions with epidemiological observations. The most serious of these relate to the level of statistical precision available from most studies. While an epidemiological study can detect a significantly increased risk of cancer with relatively few cases (provided the risk is large enough), it is another matter altogether to be able to define accurately more subtle effects, e.g., the relationship between age at first exposure and cancer risk, and whether two carcinogens act additively, multiplicatively, or otherwise. For example, consider a cohort study of lung cancer in an occupational context. A twofold increase in risk stands a 90% chance of being detected at the 5% level of significance, when there are about 13 cases of lung cancer expected on the basis of rates in a general population comparison group. Now suppose these cases are divided into two groups of equal size, according to the age employment, and hence exposure, began. A relative difference of about tenfold between the risk in the two groups would then be required for them to stand a 90% chance of being detected as significantly different from each other at the 5% level. As the number of effects being estimated increases, the level of precision available for their estimation from a study decreases. Probably for this reason, analyses of epidemiological data in the context of the multistage model have usually been descriptive in nature and not subjected to the stringency of hypothesis testing and other critical forms of statistical model fitting.

Another methodological problem confronting the epidemiologist wishing to apply the multistage model to observational data is that of confounding. Each of the model predictions described in the previous section were obtained under the condition that other factors influencing cancer risk are constant. Thus, for example, the prediction that individuals will suffer a higher excess cancer risk if exposed to a late-stage carcinogen at a later age applies under the condition that the level and temporal pattern of exposure are not related to age at exposure. In reality, when we compare groups of individuals first exposed to a carcinogen at different ages, they will generally have been exposed at a variety of levels, for differing durations, and to carcinogens in addition to those under study. Confounding arises when these additional factors are related to the variable of interest as well as the risk of disease. Thus, workers exposed to an occupational carcinogen at a young age may also be those employed during earlier calendar periods and, for the longest duration, they may have been preferentially employed in areas where higher exposure occurred or they may have been heavier smokers. There is a number of statistical solutions to the problem of confounding. The simplest is to use the confounding variables to categorize individuals and study the effect of the variables of interest within each category. Thus, in the example given above, people could be grouped according to the calendar period in which they were first employed, and the effect of age at employment on cancer risk examined within groups. A more sophisticated approach is to estimate simultaneously the relationship between risk and both the variables of interest and the confounding variables using a general mathematical formulation such as the proportional hazards model[54] or that provided by the multistage model. This article is not the place to discuss the advantages and disadvantages of the various approaches to the problem of confounding. It should, however, be stated that no solution is fully satisfactory. The subject is considered in detail in a number of epidemiological texts.[56,57]

C. Cancer Risk Following Specific Carcinogenic Exposures: Interpretation of Epidemiological Observations under the Multistage Model

The statistical problems outlined in the previous section have not discouraged a number of authors from reexamining epidemiological data from studies involving specific known carcinogens and interpreting them under the multistage model. In this section, we consider these agents in turn and discuss what is known about their effect on cancer risk.

1. Cigarette Smoking

As mentioned in Section V, the relationship between cigarette smoking and lung cancer has played a special role in cancer epidemiology, and, in particular, it has served as something of a paradigm for application of the multistage model. In a series of papers[12,14,51] the epidemiological data on lung cancer and smoking have been analyzed in great detail, producing a number of critical observations relating to the model. First, when attention is restricted to men in the British doctors study who reported no change in their smoking habits, lung cancer mortality (which is very close to incidence) in smokers increased at a much faster rate with age than among nonsmokers.[12] Figure 3 shows the age mortality curves for the two groups on a log-log plot. The curve for smokers has been adjusted for different levels of daily consumption, which may have varied with age in the group as a whole. If late-stage transitions were influenced by the exposure to cigarette smoke, we would expect rising mortality soon after exposure began (generally before age 20), becoming parallel to the line for nonsmokers after 1 or 2 decades. However, the curve for smokers seems to rise gradually at first, but then keeps rising at an increasing rate 40 to 50 years after smoking commenced. On this basis, Doll[12] concluded that an early transition rate was being increased by the exposure to tobacco smoke. When the same mortality figures are plotted on a log-log scale against time since cigarette smoking began, the third line of Figure 3 is obtained. It has a slope very close to that of the line for nonsmokers (between four and five). This finding has been interpreted as confirmation that the number of stages required for tobacco smoke carcinogenesis is between five and six, with the difference in intercept being due to the increased number of early-stage transitions produced by the tobacco smoke. Early-stage carcinogens are also defined by the independence of the excess cancer risk and age at exposure (see Section IV.A). Indeed, according to data from the U.S. veterans study[45] quoted by Day,[58] men who start smoking before age 15 have the same risk of dying from lung cancer in the age range 55 to 64 as do men who started smoking between ages 20 and 24 in the age range 65 to 74. There is a similar concordance for men who started smoking between ages 15 and 19 and after age 25, when viewed 40 to 49 years after starting smoking. However, this neat picture of an early-stage carcinogen is not supported by the evidence from the ex-smokers in the British doctors study, whose relative risk of dying of lung cancer compared to nonsmokers drops soon after cessation of smoking, while the absolute excess risk stays constant at the level attained when smoking ceased. This behavior is that predicted for an agent which increases the risk of transition to the penultimate stage before malignancy. It would be consistent with that of a single-stage carcinogen if there were but two transitions, for then the first and penultimate stages would be one and the same. However, epidemiologists have preferred to view the data as indicating that at least two stages are affected, because of the power relationship suggesting a much steeper slope than one (i.e., one less than two) for the age-incidence curve. In addition, the dose-response

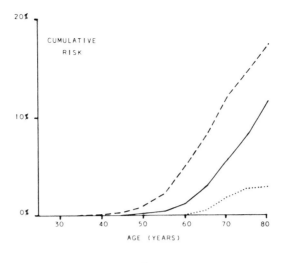

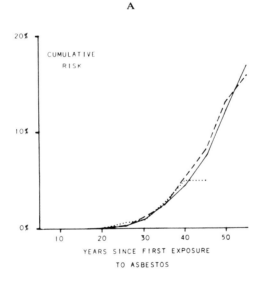

FIGURE 13. Cumulative risk of dying of mesothelioma in the absence of other causes of death among North American insulation workers first exposed to asbestos at age 18—24 (- - -), 25—34 (—), or over 35 (· · · ·). A — Plotted against age. B — Plotted against years since first exposure. (From Peto, J., Seidman, H., and Selikoff, I. J., *Br. J. Cancer*, 45, 124, 1982. With permission.)

where M is annual mesothelioma mortality, t is time since exposure began, and $b = 4.4 \times 10^{-8}$. The fit is impressive, but the fact that it was necessary to base the estimation on a restricted group of workers rather than using the whole cohort illustrates the difficulties of fitting models of this kind, even in this otherwise straightforward situation. In fact, the calendar period of first exposure could have been taken into account by allowing the constant b to depend on period. The effect of duration of exposure could also have been incorporated in this way.

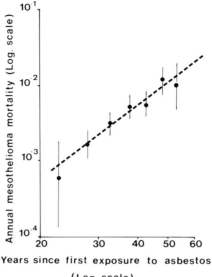

FIGURE 14. Mesothelioma mortality among North American insulation workers first exposed 1922—1946, plotted against years since first exposure. Bars indicate 95% confidence intervals (From Peto, J., Seidman, H., and Selikoff, I. J., *Br. J. Cancer,* 45, 124, 1982. With permission.)

Although Peto et al.[17] conclude on the basis of these analyses that the effect of asbestos is at an early stage of mesothelioma induction, there is less certainty regarding the exponent k. In fact, to account for a possible 10-year lag between tumor induction and death, a model of the form $M = b \cdot (t - 10)^2$ actually fits the data better than the relationship given earlier for M, suggesting that the number of hypothetical "stages" may be as few as three.

For bronchial carcinoma, the interpretation of epidemiological data is far more complex because of the important role of cigarette smoking. During the time when the men under study were being followed up, there were tremendous increases in the incidence of lung cancer in western countries, almost certainly due to the rise in cigarette consumption. Since these increases were largely on a cohort basis, it is essential that they be accounted for in studying excess lung cancer incidence among workers exposed to asbestos. The problem is compounded by the lack (in most occupational studies) of information on smoking habits at an individual level. In the studies where such information was available, the number of lung cancer deaths in nonsmokers has been very small. In a recent review of the combined effect of smoking and asbestos on lung cancer mortality, Berry et al.[68] assembled data from the five major cohort studies to date. Only 14 deaths from lung cancer occurred among nonsmokers, as compared to 426 among ex- and current smokers. Table 1 summarizes the findings reported. It was estimated that overall, the relative risk due to asbestos is about 1.8 times higher among nonsmokers than among smokers. This is in contrast to the findings of previous authors,[69,70] who concluded that smoking and asbestos exposure have a multiplicative effect on lung cancer risk, suggesting that they act on different stages of the carcinogenesis process. Thomas[71] used data from the study on Quebec asbestos miners and

Table 1
LUNG CANCER AND SMOKING IN FIVE COHORT STUDIES OF ASBESTOS WORKERS

Study	Nonsmokers			Smokers		
	Observed	Expected	Relative risk	Observed	Expected	Relative risk
Insulation workers, N.Y. and N.J., 1963—1967	0	0.05	0	24	2.98	8.1
Insulators, U.S. and Canada, 1967—1976	4	0.7	5.7	268	51.0	5.3
Amosite factory workers, 1961—1977	5	0.2	25.0	45	9.6	4.7
Factory workers (women), U.K., 1971—1980	1	0.2	5.0	14	1.9	7.4
Factory workers, U.K., 1971—1980	4	0.55	7.3	75	31.02	2.4

From Berry, G., Newhouse, M. L., and Antonis, P., *Br. J. Ind. Med.*, 42, 12, 1985. With permission.

millers[72] to estimate the parameters of the multistage model, and obtained the result that under the assumption that bronchial carcinogenesis involves six stages, asbestos acts at the fourth, and smoking at the fifth. This work probably represents the most ambitious effort yet undertaken to apply the multistage model to epidemiological data. Statistical criteria for model fitting rather than heuristic interpretation of incidence patterns are used to estimate the stages at which the two agents act, taking into account the level and duration of exposure. It remains to be seen whether the results can be replicated on other data sets, or if they simply serve as sophisticated descriptive indexes applicable only to the lung cancer risk in the Quebec asbestos industry.

Other information on lung cancer risk following asbestos exposure is provided by a study in which the majority of workers were exposed for a short time only (less than 2 years) in an asbestos factory in New Jersey.[73] Walker[74] used the results of this study to calculate the risk of lung cancer in relation to time since exposure, and concluded that compared to expected mortality, the relative risk rises to a peak about 20 years after exposure, and then declines. This observation could be consistent with an excess risk which first increases rapidly with age and then stabilizes, as one would expect after cessation of exposure to a late-stage carcinogen. However, it could also occur if the excess risk continued to rise, albeit at a slower rate than the background. The picture is muddied further by the suggestion that asbestos fibers may remain in the lung and exert a carcinogenic effect long after exposure has ceased.[75]

There is relatively little information available on the effect on lung cancer incidence of the age at which exposure to asbestos occurred. However, among the workers studied by Seidman et al.,[73] the excess probability of lung cancer appears to increase with age at exposure. Table 2 gives the observed and expected cumulative probabilities of dying of lung cancer, by age at start of exposure, months employed, and years since exposure began. These probabilities are roughly the summation of the incidence over 10-year time periods since exposure began. For all combinations of months employed

Table 2
EXCESS PROBABILITY OF LUNG CANCER FOLLOWING SHORT-TERM EXPOSURE TO ASBESTOS

Duration of employment	Age at first exposure	Years since first exposure		
		5	15	25
<9 months	15—24	−0.02		
	25—34	−0.11	0.76	
	35—44	1.40	1.96	1.21
	45—54	−1.26	3.09	5.76
≥9 months	15—24	−0.02		
	25—34	−0.10	2.53	
	35—44	5.25	5.85	8.61
	45—54	11.07	7.21	11.61

From Seidman, H., Selikoff, I. J., and Hammond, E. C., *Ann. N.Y. Acad. Sci.,* 330, 61, 1979. With permission.

and years since employment began, the excess probability increases steadily with the age at which employment occurred. This table also complements the relative risk calculations reported above from the same study, and gives some indication of a roughly constant excess risk following the exposure.

Observation of cancer risk in workers who have been exposed to asbestos illustrates well the difficulties in drawing mechanistic conclusions based on the multistage model. While asbestos appears to initiate the process of mesotheliomagenesis, its effect on lung carcinogenesis seems to be tied up with age, smoking habits, and possibly other factors.

4. Nickel

Exposures occurring during the process of nickel refining have been associated with the induction of cancers in the nasal sinuses and lungs. The most informative study of the temporal effects of exposure in these workers has been in a South Wales refinery,[76] where 56 men with cancer of the nasal sinuses and 137 with lung cancer have now been identified. These numbers are, respectively, 50 and 5.6 times the expectation based on national rates. The responsible agents have never been specifically identified and may never be, since the refining process and working conditions during the high risk years earlier this century have changed radically. Thus one must rely on surrogate measures of exposure level, i.e., the calendar period and the area in the refinery in which the individuals worked.

Like mesothelioma, carcinoma of the nasal sinuses is a very rare tumor which has nonetheless been associated with a number of occupational exposures in addition to those occurring during nickel refining. One might, therefore, suspect that it was initiated by these exposures and show similar age independence to the risk of mesothelioma following asbestos exposure (see previous section). However, the observation on the South Wales workers suggest otherwise. Table 3 gives the estimate of excess risk compared to the general population of England and Wales associated with increasing ages of commencing employment and time since employment began. The estimates are adjusted for each other, the duration of exposure in high risk areas of the refinery, and the calendar period of first employment.[20] The excess risk increases rapidly with age at

Table 3
ADJUSTED EXCESS RISK OF NASAL SINUS AND
LUNG CANCER AMONG SOUTH WALES NICKEL
REFINERY WORKERS BY AGE AT FIRST
EXPOSURE AND TIME SINCE FIRST EXPOSURE

	Lung cancer		Nasal sinus cancer	
	Number of cases	Excess risk[a]	Number of cases	Excess risk[a]
Age at first exposure				
<20	13	1.0	2	1.0
20—27.5	72	2.8[b]	20	5.1[b]
27.5—35	41	2.9[b]	20	10.9[c]
35+	11	2.6	14	36.6[c]
Time since first exposure (years)				
≤20	6	1.0	1	1.0
20—30	35	2.8[b]	19	4.9[b]
30—40	55	4.9[c]	17	6.6[b]
40—50	31	4.8[c]	13	13.9[c]
50+	10	2.4	6	24.7[c]

[a] Relative to that in the baseline category, which has value 1.0.
[b] Significantly different from the baseline at the 0.05 level of significance.
[c] Significantly different from the baseline at the 0.01 level of significance.

From Kaldor, J., Peto, J., Easton, D., Doll, R., Hermon, C., and Morgan, L., *J. Natl. Cancer Inst.*, October, 1986. With permission.

exposure, but also with time since exposure began. In fact, the relative risk compared to the general population also increases rapidly with age at exposure (Table 3). This is surprising, since even for a late-stage carcinogen, the number of cells ready to make the final transition to malignancy could not increase with age at a faster rate than cancer incidence, as is implied by this finding. However, if one assumes that the effect of exposure is to increase the exponential proliferation of late-stage cells,[21,22] an effect of this kind could be predicted to occur.[124] The relative risk appears to have been rather constant from 20 years after employment began, when in fact exposure for most workers would have ceased due to radical changes in the refining process and industrial hygiene.[77] Since a constant relative (and rising excess) risk following cessation of exposure suggests an early-stage carcinogenic effect, employment in the refinery seems to exhibit both early- and late-stage effects on nasal sinus carcinogenesis.

Interpretation of the lung cancer results is again complicated by the large cohort-related increase in rates among the population as a whole. Carrying out analyses of the kind described above for nasal sinus cancer, again reported in Table 3, it was found that age at exposure had little influence on the relative risk. There was, however, an increase in excess risk, although far less dramatic than that seen for nasal sinus cancer. The excess risk also rose steadily following the start of employment, but then plateaued or dropped after about 40 years. Thus, late-stage effects are again suggested, but not particularly clearly. The most striking feature of the lung cancer results is the rapid drop in the relative risk compared to the general population, with time since employment began. While this could result from cessation of exposure to a late-stage carcinogen, it could also occur for an early-stage agent if another early-stage agent, (e.g., tobacco smoke) was acting on underlying population rates to produce an excess rising much faster than the excess due to nickel exposure. One small cohort study in Norway suggests that cigarette smoking and nickel refining combine additively in their effect on lung cancer risk,[78] further supporting this idea. An alternative interpretation is that those most susceptible to pulmonary nickel carcinogenesis (possibly the smokers) died early, leaving men who were at lower (relative) risk of dying of lung cancer.[76,79]

5. Arsenic

Although a number of toxic exposures has been known to occur during the smelting of copper, the excess of lung cancer seen among smelter workers has generally been attributed to inorganic arsenic compounds.[80] Brown and Chu[18] have reanalyzed data from the study of Montana smelter workers[81] among whom 278 lung cancer deaths have been recorded, just over twice the number expected on the basis of national death rates. Their analysis was carried out with specific reference to the Armitage-Doll multistage model. Taking into account estimated exposure level and duration of exposure, the excess risk of lung cancer increased with age at exposure, but was unrelated to time since stopping exposure when analysis was restricted to men terminating work after age 55 (to minimize the number of men occupationally exposed elsewhere after leaving the smelter). These two relationships are consistent with late-stage carcinogenesis. As a by-product of the analysis, the exponent relating age to risk was estimated, resulting in a value of 5.8, which corresponds closely to the estimate of 5.6 obtained as the exponent (i.e., the slope of the log-log plot) for the age-lung cancer mortality relationship for the period 1940 to 1965.

As was the case for asbestos and nickel exposure, the absence of smoking information on an individual level makes any mechanistic inferences about lung cancer risk somewhat tentative.

Some information on the interaction between smoking and arsenic exposure in the induction of lung cancer is provided by a case-control study in a cohort of copper smelter workers in Sweden.[82] After adjusting for the effect of age, it appeared that the combined effect was closer to multiplicative than additive, suggesting that different stages are being affected by the two exposures.

6. Aromatic Amines

The fact that cancer of the bladder can result from exposure to certain aromatic amines used in the synthetic dye industry has been recognized since the end of the 19th century.[83-85] A number of studies in Europe, North America, and Japan reported phenomenally high rates of bladder cancer among workers exposed to β-naphthylamine, benzidine, 4-aminobiphenyl, and other compounds. For example, in the group of workers reported on by Goldwater et al.,[86] 96 of 366 men were diagnosed with bladder cancer, while among 25 workers studied by Zavon et al.,[87] 13 developed bladder tumors.

Nevertheless, most of these studies have been relatively uninformative with regard to the evolution of risk with time and the effect of age at exposure on risk. Even if the information was presented in a suitable manner, the interpretation of a number of the more recent studies would be complicated by the fact that most cases were detected at single cytological surveys of worker groups, rather than at incidence or death.

One study which has been analyzed with specific regard to the multistage model is that of dyestuff workers in a factor in Turin.[88] Of 906 men employed in the factory, 41 have died of bladder cancer, representing a risk 46 times higher than that of the Italian male population. Using a multivariate analysis similar to those described above for the nickel- and arsenic-exposed workers, it was found that the excess risk (very close to the absolute) of bladder cancer did not depend on age at exposure when the duration of exposure and time since exposure were taken into account. In addition, the excess continued to rise for at least 10 years following cessation of exposure, while the relative risk declined somewhat. Apart from this last observation, the results are consistent with a carcinogen acting at an early stage. In fact, in another study, the relative risk appeared to rise over 50 years after cessation of employment in occupations hazardous for bladder cancer.[89]

7. Other Industrial Exposures

Various other occupational exposures have been found to be carcinogenic. *Bis*-chloromethyl ether has been shown to be responsible for lung cancer of the small or oat cell type in several studies,[84] as have various exposures occurring during the production of chromium or chromate products.[80] Leukemia has been observed at elevated rates in workers exposed to benzene,[85] and angiosarcoma can result following vinyl chloride exposure.[90] However, the epidemiological studies which described these associations have not reported analyses which could be used to elucidate the mechanisms of carcinogenesis for these exposures, even in the tentative way described above. For some of the rarer cancers, the number of cases described is simply too small to permit detailed analysis, while in other studies, breakdowns of risk according to variables of interest have not been provided. There is some suggestion that *bis*-chloromethyl ether acts at an early stage of carcinogenesis, since the relative risk of lung cancer appears to remain constant after exposure has ceased.[91] More detailed analyses of these data would be required before any more firm conclusions can be drawn.

8. Hormones

It is clear that both endogenous and exogenous hormones can play a role in human carcinogenesis. Although there is evidence of their involvement for a number of tissues, they have been most strongly implicated for cancers of the breast and the endometrium.[92,93] In the following, we therefore restrict attention to these two sites. For detailed references, see the reviews by de Waard[94] for the endometrium, Petrakis et al.[95] for the breast, and Armstrong et al.[92] and Henderson et al.[96] for endogenous hormones and carcinogenesis in general.

It is believed that estrogens are the most important hormonal factor in the etiology of both endometrial and breast cancer. The two cancers certainly have a number of epidemiological features in common, including a reduced risk associated with early childbearing and high parity and an increased risk for women with late menopause, which suggests the common influence of some aspect of ovarian function. However, there are major differences which indicate that more complex, and as yet unexplained, mechanisms may be operating. The most basic is that while breast cancer incidence increases continuously with age, apart from the inflexion around the time of menopause (see Figure 4), the endometrial cancer age-incidence curve plateaus around the menopause, and in some populations actually drops in late middle age.[36] In addition, and perhaps in contrast, the endometrium seems to be far more sensitive than the breast to the effect of postmenopausal replacement estrogens and the increased levels of extraovarian estrogens associated with obesity.

Although the most significant source of estrogen exposure in the lifetime of most women is endogenous, one can view the exposure history in much the same way as the agents we have considered earlier. Exposure to ovarian estrogens commences sometime before menarche and continues in accordance with the menstrual cycle until the menopause. Ovarian estrogen production effectively ceases at the menopause, whether natural or induced, although there may be appreciable levels associated with obesity. During pregnancy, the placenta produces increasing amounts of estrogens and soon becomes the overwhelming site of hormone production in the body. Added to these endogenous exposures may have been exogenous estrogens, taken as oral contraceptives, postmenopausal replacement, or for other therapeutic reasons.

From the viewpoint of the multistage model, circumstances resulting in large changes in exposure would be of primary interest. These include menarche, full-term pregnancy, natural or induced menopause, and the withdrawal of replacement therapy. While it is clear that the timing of menarche and pregnancies in the lifetime of a woman plays a crucial role in determining her risk for developing breast cancer and probably endometrial cancer, these factors have not been studied in sufficient detail to attempt analyses of the kind required. For example, it would be of interest to know whether the risk of breast cancer for a woman is determined simply by the amount of time since the onset of menarche, or if total age was also important. If estrogen acts at a late stage, one would predict the latter and, in addition, that the risk would increase rapidly, albeit for a short time, following the massive increase in estrogen levels during pregnancy. Unfortunately, data which would address these questions are as yet unavailable. There is, however, more information concerning the effects of menopause and replacement therapy withdrawal. We consider these factors in turn.

In previous sections, we have compared the cancer risk in those exposed to various agents to that in unexposed individuals. For estrogens, an unexposed group of women does not exist. We must therefore relate the risk in those for whom exposure ceases, to the risk among those who continue to be exposed. If an agent was responsible for early stage transition, the risk would continue rising with age just as it had during exposure for a long time after cessation. If a late stage was affected, we would expect a relative flattening of the curve, much as is in fact seen for breast cancer incidence around the time of the menopause. Under the two-stage model,[21] similar effects would be predicted if the cellular proliferation rates were affected by the agent.[48] For endometrial cancer, incidence effectively freezes around the late forties, and it would therefore appear that the number of penultimate stage cells does not increase during this time and that the ovarian estrogens are responsible for the overwhelming majority of transitions to this stage. Of course these are conclusions drawn from population age-incidence curves which reflect the experience of women who undergo menopause across a wide age range. For breast cancer, they are further supported by individual observation of women who have had an artificially induced menopause. Figure 15 shows the subsequent evolution of relative risk for breast cancer as compared to the general population among women who underwent surgical oophorectomy[97] or who probably had a menopause induced by high doses of radiation.[98] In both studies, the relative risk of breast cancer was halved within a few years of the treatment. Since oophorectomy is generally accompanied by hysterectomy, it is not possible to examine the risk of endometrial cancer in women undergoing surgical sterilization. However, Figure 15 also gives the evolution of the risk among women who received radiation therapy for cervical cancer. The numbers of cases are small, but the relative risk does not seem to change for at least 15 years following therapy. The possibility that an appreciable percentage of these women underwent hysterectomy after the radiotherapy should be noted.

The association between endometrial cancer and the use of conjugated estrogens to attenuate symptoms of the menopause was established by a number of epidemiological studies in the latter part of the 1970s,[93] by which time many middle-aged women, particularly in the U.S., had been exposed. There is also some evidence that replacement estrogens increase the risk of breast cancer in postmenopausal women.[99-101] Earlier studies of estrogen use suggested that the risk of endometrial cancer declines ex-

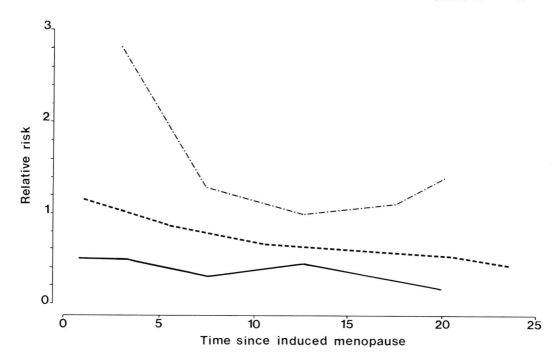

FIGURE 15. Relative risk of breast and endometrial cancer following artificial menopause. ——— — Breast cancer following irradiation for cervical cancer. - - - - — Breast cancer following surgical oophorectomy. -·-·- — Endometrial cancer following irradiation for cervical cancer.

tremely rapidly following discontinuation of exposure. Jick et al.[102] studied women in a group health plan in the U.S. They estimated the incidence of endometrial cancer among current users of conjugated estrogens (defined as use in the 6-month period before diagnosis in cases) and nonusers aged 50 to 64 by 6-month calendar periods in 1975 and 1976. The tenfold greater risk in the former group has been taken as evidence that the risk of endometrial cancer drops to background levels within 6 months after drug use stops. If this is the case, the effect of replacement estrogens would appear to be on the last stage before malignancy. However, it is not clear that the low risk in the nonusers even derives from women who were medium or long-term users in the past, since the study was not designed to elicit information of this kind. In fact, in a much larger, more recent study,[103] the risk of endometrial cancer appeared well in excess more than 10 years after last estrogen use, especially for women who had taken the drugs for more than 1 year. As shown in Figure 16, there was no clear evidence of a decline in risk during the period after exposure. One might be convinced that the relative risk remained roughly constant, indicating a penultimate stage effect, since the background and hence the excess are also constant with respect to age in the period under consideration.

V. THE MULTISTAGE MODEL AND PRECURSOR STATES

Explicit specification of the biological meaning of a stage has been scrupulously avoided by most previous authors who have discussed epidemiological observations in the context of the Armitage-Doll multistage model. Indeed, although a number of cancers was known to be associated with morphologically well-defined precursor lesions, the means for genetically distinguishing normal, precursor, and malignant cells had

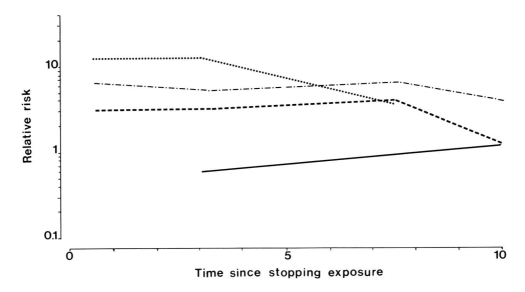

FIGURE 16. Relative risk of endometrial cancer following cessation of estrogen replacement therapy after a duration of <1 year (———), 1—4 years (- - - -), 5—9 years (-·-··-), or ≥10 years (· · · ·) of usage.

not been adequately developed. However, as developments in cytogenetics and molecular genetics over the last decade have brought us closer to this possibility, interest has focused on the precise nature of the cells which are precursors to malignancy. Moolgavkar and Knudson,[21] in their presentation of the two-stage model of carcinogenesis, suggested that the two events in the pathway to malignancy were simply mutations at the same site in homologous chromosomes in a gene which would now be referred to as an oncogene (or an "antioncogene"[104]) for a given cancer. This theory accounts for two important observations in cancer epidemiology. The first is the familial clustering of specific cancers, which is explained as occurring through the inheritance of the mutated gene on one chromosome. Families in which this gene is present are at higher risk for the cancer, because a somatic mutation in the homologous chromosome of a single cell can give rise to malignant transformation of the cell. For nonfamilial cases, a somatic mutation is required in both chromosomes. The second observation concerns precursor lesions, of which for certain cancers there seem to be two kinds, one familial and one sporadic. For example, carcinomas of the colon appear to occur in adenomatous polyps, which in turn may arise sporadically in the colon or extensively in individuals with familial polyposis. Such people are believed to have a lifetime risk for carcinoma of the colon which is close to 100%.[105] According to the two-stage theory, the polyps arise from cells which have already undergone the first transition, and all the cells of polyposis patients are in this state.

Another example is provided by melanoma, which is believed to be preceded by skin lesions known as nevi. These occur either sporadically or arise in the dysplastic nevus syndrome. Individuals with this syndrome, which like familial polyposis appear to be inherited as an autosomal dominant trait, are at an enormously increased risk of melanoma, and again, it has been suggested that their lifetime cumulative incidence approaches 100%.[106]

The model proposed by Moolgavkar and Knudson[21] was developed from earlier observations by Knudson[107] on retinoblastoma, which has since been demonstrated to be associated with exactly the kind of homozygosity predicted by the model.[108,109] There have been similar findings for Wilms' tumor.[110-112] In both these cases, the cancer gene appears to be recessive, and the heterozygotic cells are phenotypically normal. This suggests an oncogenic mechanism somewhat different to that operating for carcinoma of the colon, melanoma, and the other cancers for which there seem to be phenotypically well-defined precursors.

Another cancer which has been well-characterized genetically is Burkitt's lymphoma. In contrast to the tumors mentioned above, a heritable susceptibility has not been established. However, at least in the African form of the disease, at least two distinct stages can be ascribed to its natural history. The first is the infection of target B-cells by Epstein-Barr virus, which causes a massive proliferation and possibly a heritable change.[113,114] The second is the chromosomal translocation which appears to result in activation of the *c-myc* oncogene.[115] Whether other stages are also involved and whether there is an obligatory temporal sequence of transitions between stages are not clear.

An interesting contrast is provided by chronic myelocytic leukemia, which is also associated with a chromosomal translocation.[116] However, the translocation has been detected in otherwise normal stem cells, whereas normal lymphocytes appear never to exhibit the translocation associated with Burkitt's lymphoma.

For the vast majority of human cancers, information of this kind is still lacking. Oncogenes have been detected in tumor cells for a number of cancers,[117] but never with a frequency approaching the 100% seen for translocation in African Burkitt's lymphoma. It is, of course, possible that for these cancers, multiple oncogenes can be involved and that not every case of malignant disease will obligatorily present the same genetic alterations. For those cancers for which the pathological natural history has been well-characterized, there is potentially the opportunity to study oncogenic activation in precursor as well as fully malignant cells. For example, biopsy samples from the large number of women systematically detected with cervical dysplasia and carcinoma *in situ* through screening programs would provide very useful material for research of this kind. The peak incidence of invasive carcinoma occurs long after that of carcinoma *in situ* detection,[118] suggesting that further heritable changes may be occurring. In the case of cervical cancer, an important question currently being investigated is the importance of the human papilloma virus (HPV) in its etiology.[119] Of particular interest, from the perspective of the multistage model, is the relationship between its incorporation into the cellular genome and the stage of cervical abnormality. A recent survey by Schneider et al.[120] found that the HPV subtype most associated with invasive cancer was also detected with a high frequency in cases of dysplasia and carcinoma *in situ*. The sample sizes in this study were too small to draw firm conclusions with regard to the comparative rates of HPV incorporation among the different precursor lesions.

VI. CONCLUSIONS

In this paper, we have tried to critically review epidemiological data which relate to the various multistage models for carcinogenesis. It has not been our goal to consider multistage carcinogenesis in general, but rather to evaluate the evidence arising purely from observations in humans.

Originally, the models were used simply as a device to predict the increasing incidence of cancer with age. It has often been pointed out that many diseases, most obviously those of the cardiovascular system, become more common in older people, and yet it has not been thought appropriate to utilize a multistage formulation for the disease process. However, the fact that so many agents of diverse types seemed capable of inducing cancer in humans led to a need to attempt to classify them in some way, and the multistage model offered itself as the natural framework in which to do this.

By now, the epidemiological literature has been fairly well ransacked for usable data, and we are probably at or beyond the limits of observational precision to detect the effects that interest us. It is therefore appropriate to examine what has been learned about the biology of cancer from interpreting epidemiological data in the ways described in this paper.

There are certain aspects of carcinogenesis which are indisputable. Cancer cells are different from normal cells, and their progeny differ in the same way, so they must have gone through a heritable change (or changes) to become malignant. The key additional aspects of the multistage formulation are, first, that there is more than one transition required and, second, that different agents can affect different stages. Human observation has provided convincing evidence in support of the first claim. A single mutation model would predict a constant age-incidence curve provided the number of susceptible normal cells remained constant, and extreme hypotheses relating this number to age per se would then be required to account for the increases in cancer incidence seen with age.

The second proposition remains less clearly supported when one restricts attention to human data. As we saw, very few agents induced cancer in a manner fully consistent with the predictions for an early-stage agent. Only the one study of asbestos-induced mesothelioma demonstrated a risk which appeared both to be independent of age at exposure and to continue rising long after exposure ceased. It is possible that most agents which are capable of initiating cancer also increase the rate of later transitions, so that pure early-stage agents would be rarely observed. Cigarette smoking, radiation, and nickel refining may be in this category. Moreover, it is impossible to distinguish whether an age-related increase in risk is due to a late-stage or promoter effect or to an entirely different mechanism, i.e., the decline in immune surveillance proposed by Burnet[121] and so strongly criticized by proponents of the multistage model.[26,27] Of course, when we look outside the epidemiological literature, clear distinctions between carcinogens emerge on the basis of their genotoxic activity. For example, estrogens consistently appear to have late-stage effects in human studies and, indeed, have never been found to have any activity in short-term tests for genetic damage. Asbestos, the most clearly defined early-stage agent, appeared to contradict this reasoning by also being neutral in many such assays until it was recently found to cause chromosomal aberrations and aneuploidy in vitro.[122,123]

At the very least, the scrutiny of epidemiological data by statisticians and others looking for support for mathematical models of carcinogenesis has shown us that a simplistic model is *not* tenable. It has also resulted in improvement in data collection and presentation and provided the impetus for experimental work in carcinogenesis mechanisms. It remains to be seen whether epidemiological findings will have more to say about the fundamental mechanisms of carcinogenesis or whether the solutions to these problems lie in the hands of molecular biologists and others working at the other end of the spectrum of cancer research.

REFERENCES

1. Berenblum, I. and Shubik, P., A new, quantitative, approach to the study of the stages of chemical carcinogenesis in the mouse's skin, *Br. J. Cancer,* 1, 383, 1947.
2. Mottram, J. C., A developing factor in experimental blastogenesis, *J. Pathol. Bacteriol.,* 56, 181, 1944.

3. Slaga, T. J., Fischer, S. M., Nelson, K., and Gleason, G. L., Studies on the mechanism of skin tumor promotion: evidence for several stages in promotion, *Proc. Natl. Acad. Sci. U.S.A.*, 77, 3659, 1980.
4. Hennings, H., Shores, R., Wenk, M. L., Spangler, E. F., Tarone, R., and Yuspa, S. H., Malignant conversion of mouse skin tumors is increased by tumor initiators and unaffected by tumor promoters, *Nature (London)*, 304, 67, 1983.
5. Peraino, C., Fry, R. J. M., and Staffeldt, E., Enhancement of spontaneous hepatic tumorigenesis in C3H mice by dietary phenobarbital, *Cancer Res.*, 31, 1506, 1971.
6. Fukushima, S., Imaida, K., Sakata, T., Okamura, T., Shibata, M., and Ito, N., Promoting effects of sodium L-ascorbate on 2-stage urinary bladder carcinogenesis in rats, *Cancer Res.*, 43, 4454, 1983.
7. Farber, E. and Cameron, R., The sequential analysis of cancer development, *Adv. Cancer Res.*, 31, 125, 1980.
8. Farber, E., The multistep nature of cancer development, *Cancer Res.*, 44, 4217, 1984.
9. Schmähl, D., Critical remarks on the validity of promoting effects in human carcinogenesis, *J. Cancer Res. Clin. Oncol.*, 109, 260, 1985.
10. Williams, T. R. and Clark, W. L., Science and politics: the possible regulation of cancer promoters, *Environ. Health Perspect.*, 50, 351, 1983.
11. Nordling, C. O., A new theory on the cancer inducing mechanism, *Br. J. Cancer*, 7, 68, 1953.
12. Doll, R., The age distribution of cancer: implications for models of carcinogenesis, *J. R. Stat. Soc. A*, 134, 133, 1971.
13. Cook, P. J., Doll, R., and Fellingham, S. A., A mathematical model for the age distribution of cancer in man, *Int. J. Cancer*, 4, 93, 1969.
14. Doll, R. and Peto, R., Cigarette smoking and bronchial carcinoma: dose and time relationships among regular smokers and life-long non-smokers, *J. Epidemiol. Commun. Health*, 32, 303, 1978.
15. Day, N. E. and Brown, C. C., Multistage models and the primary prevention of cancer, *J. Natl. Cancer Inst.*, 64, 977, 1980.
16. Whittemore, A. S., The age distribution of human cancer for carcinogenic exposures of varying intensity, *Am. J. Epidemiol.*, 106, 418, 1977.
17. Peto, J., Seidman, H., and Selikoff, I. J., Mesothelioma mortality in asbestos workers: implications for models of carcinogenesis and risk assessment, *Br. J. Cancer*, 45, 124, 1982.
18. Brown, C. C. and Chu, K. C., Implications of the multistage theory of carcinogenesis applied to occupational arsenic exposure, *J. Natl. Cancer Inst.*, 70, 455, 1983.
19. Peto, J., Cuckle, H., Doll, R., Hermon, C., and Morgan, L. C., Respiratory cancer mortality of Welsh nickel refinery workers, in *IARC Scientific Publications No. 53, Nickel in the Human Environment*, Sunderman, F. W., Jr., Ed., International Agency for Research on Cancer, Lyon, 1984, 37.
20. Kaldor, J., Peto, J., Easton, D., Doll, R., Hermon, C., and Morgan, L., Models for respiratory cancer in nickel refinery workers, *J. Natl. Cancer Inst.*, October, 1986.
21. Moolgavkar, S. H. and Knudson, A. G., Jr., Mutation and cancer: a model for human carcinogenesis, *J. Natl. Cancer Inst.*, 66, 1037, 1981.
22. Moolgavkar, S. H., Model for human carcinogenesis: action of environmental agents, *Environ. Health Perspect.*, 50, 285, 1983.
23. Peto, J., Early- and late-stage carcinogenesis in mouse skin and in man, in *IARC Scientific Publications No. 56, Models, Mechanisms and Etiology of Tumour Promotion*, Börszönyi, M., Day, N. E., Lapis, K., and Yamasaki, H., Eds., International Agency for Research on Cancer, Lyon, 1984, 359.
24. Armitage, P. and Doll, R., The age distribution of cancer and a multistage theory of carcinogenesis, *Br. J. Cancer*, 8, 1, 1954.
25. Armitage, P. and Doll, R., Stochastic models for carcinogenesis, in *Proc. 4th Berkeley Symp. on Mathematical Statistics and Probability: Biology and Problems of Health*, University of California Press, Berkeley, 1961, 19.
26. Peto, R., Epidemiology, multistage models and short-term mutagenicity tests, in *Origins of Human Cancer*, Hiatt, H. H., Watson, J. D., and Winsten, J. A., Eds., Cold Spring Harbor Laboratory, New York, 1977, 1403.
27. Peto, R., Parish, S. E., and Gray, R. G., There is no such thing as ageing, and cancer is not related to it, in *IARC Scientific Publications No. 58, Age-Related Factors in Carcinogenesis*, Likhachev, A., Anisimov, V., and Montesano, R., Eds., International Agency for Research on Cancer, Lyon, 1985, 43.
28. Moolgavkar, S. H., The multistage theory of carcinogenesis and the age distribution of cancer in man, *J. Natl. Cancer Inst.*, 61, 49, 1978.
29. Pike, M. C., A method of analysis of a certain class of experiments in carcinogenesis, *Biometrics*, 22, 142, 1966.
30. Armitage, P., Multistage models of carcinogenesis, in *Environ. Health Perspect.*, 63, 195, 1985.

31. Moolgavkar, S. H. and Venzon, D. J., Two-event models for carcinogenesis: incidence curves for childhood and adult tumors, *Math. Biosci.*, 47, 55, 1979.
32. Pike, M. C., "Hormonal" risk factors, "breast tissue age" and the age-incidence of breast cancer, *Nature (London)*, 303, 767, 1983.
33. Doll, R., Payne, P., and Waterhouse, J., Eds., *Cancer Incidence in Five Continents. A Technical Report*, International Union Against Cancer, Geneva, 1966.
34. Doll, R., Payne, P., and Waterhouse, J., Eds., *Cancer Incidence in Five Continents*, Vol. 2, International Union Against Cancer, Geneva, 1970.
35. Waterhouse, J., Muir, C., Correa, P., and Powell, J., Eds., *IARC Scientific Publications No. 15, Cancer Incidence in Five Continents*, Vol. 3, International Agency for Research on Cancer, Lyon, 1976.
36. Waterhouse, J., Muir, C., Shanmugaratnam, K., and Powell, J., Eds., *IARC Scientific Publications No. 12, Cancer Incidence in Five Continents*, Vol. 4, International Agency for Research on Cancer, Lyon, 1982.
37. Bjarnason, O., Day, N., Snaedal, G., and Tulinius, H., The effect of year of birth on the breast cancer age-incidence curve in Iceland, *Int. J. Cancer*, 13, 689, 1974.
38. Moolgavkar, S. H., Stevens, R. G., and Lee, J. A. H., Effect of age on incidence of breast cancer in females, *J. Natl. Cancer Inst.*, 62, 493, 1979.
39. Boyle, P., Day, N. E., and Magnus, K., Mathematical modelling of malignant melanoma trends in Norway, 1953—1978, *Am. J. Epidemiol.*, 118, 887, 1983.
40. Holford, T. R., The estimation of age, period and cohort effects for vital rates, *Biometrics*, 39, 311, 1983.
41. Elwood, J. M., Gallagher, R. P., Hill, J. J., Spinelli, J. J., Pearson, J. C. G., and Threlfall, W., Pigmentation and skin reaction to sun as risk factors for cutaneous melanoma: Western Canada Melanoma Study, *Br. Med. J.*, 288, 99, 1984.
42. Green, A. C., MacLennan, R., and Siskind, V., Common acquired naevi and the risk of malignant melanoma, *Int. J. Cancer*, 35, 297, 1985.
43. Holman, C. D. J. and Armstrong, B. K., Cutaneous malignant melanoma and indicators of total accumulated exposure to the sun: an analysis separating histogenetic types, *J. Natl. Cancer Inst.*, 73, 75, 1984.
44. Doll, R. and Hill, A. B., Mortality in relation to smoking: ten years' observations of British doctors, *Br. Med. J.*, 2, 1071, 1964.
45. Kahn, H., The Dorn study of smoking and mortality among US veterans: report on eight and one-half years of observation, *Natl. Cancer Inst. Monogr.*, 19, 1, 1966.
46. Hammond, E. C., Smoking in relation to the death rates of one million men and women, *Natl. Cancer Inst. Monogr.*, 19, 127, 1966.
47. Day, N. E. and Charnay, B., Time trends, cohort effects and aging as influence on cancer incidence, in *Trends in Cancer Incidence: Causes and Practical Implications*, Magnus, K., Ed., Hemisphere Press, New York, 1981, 51.
48. Moolgavkar, S., Day, N. E., and Stevens, R. G., Two-stage models for carcinogenesis: epidemiology of breast cancer in females, *J. Natl. Cancer Inst.*, 65, 559, 1980.
49. Shanmugaratnam, K., Lee, H. P., and Day, N. E., *IARC Scientific Publications No. 47, Cancer Incidence in Singapore 1968—1977*, International Agency for Research on Cancer, Lyon, 1984.
50. Wright, D. H. and Serck-Hanssen, A., Lymphoreticular tumours, in *Tumours in a Tropical Country*, Templeton, A. C., Ed., Springer-Verlag, Berlin, 1973, 270.
51. Doll, R. and Peto, R., Mortality in relation to smoking: 20 years' observations on male British doctors, *Br. Med. J.*, 2, 1525, 1976.
52. Beebe, G. W., Kato, H., and Land, C. E., Mortality Experience of Atomic Bomb Survivors, 1950—74, RERF Tech. Rep. 1-77, Hiroshima, Japan, 1977.
53. Siemiatycki, J. and Thomas, D. C., Biological models and statistical interactions: an example from multistage carcinogenesis, *Int. J. Epidemiol.*, 10, 383, 1981.
54. Breslow, N. E., Lubin, J. H., Marek, P., and Langholz, B., Multiplicative models and cohort analysis, *J. Am. Stat. Assoc.*, 78, 1, 1983.
55. Cardis, E. M., Modelling the Effect of Exposure to Environmental Carcinogens on Incidence of Cancers in Populations, Ph.D. thesis, University of Seattle, Washington, 1985.
56. Breslow, N. E. and Day, N. E., *IARC Scientific Publications No. 32, Statistical Methods in Cancer Research*, Vol. 1, International Agency for Research on Cancer, Lyon, 1980.
57. Kleinbaum, L., Kupper, L. L., and Morgenstern, H., *Epidemiologic Research: Principles and Quantitative Methods*, Lifetime Learning Publications, Belmont, Calif., 1982.
58. Day, N. E., Time as a determinant of risk in cancer epidemiology: the role of multi-stage models, *Cancer Surv.*, 2, 577, 1983.

59. Whittemore, A. S. and Altshuler, B., Lung cancer incidence in cigarette smokers: further analysis of Doll and Hill's data for British physicians, *Biometrics,* 32, 805, 1976.
60. Hoffmann, D. and Wynder, E. L., Tobacco carcinogenesis. XI. Tumor initiators, tumor acceleration and tumor promoting activity of condensate fractions, *Cancer,* 27, 848, 1971.
61. Smith, P. G., Radiation, in *Cancer Risks and Prevention,* Vessey, M. P. and Gray, M., Eds., Oxford University Press, Oxford, 1985, 119.
62. Smith, P. G. and Doll, R., Mortality among patients with ankylosing spondylitis after a single treatment course with x rays, *Br. Med. J.,* 284, 449, 1978.
63. Darby, S. C., Modelling age- and time-dependent changes in the rates of radiation-induced cancers, in *Atomic Bomb Survivor Data: Utilization and Analysis,* Prentice, R. L. and Thompson, D. J., Eds., SIAM Institute for Mathematics and Society, Philadelphia, 1984, 67.
64. Boice, J. D., et al., Second cancers following radiation treatment for cervical cancer. An international collaboration among cancer registries, *J. Natl. Cancer Inst.,* 74, 955, 1985.
65. Boice, J. D., Jr., Land, C. E., Shore, R. E., Norman, J. E., and Tokunaga, M., Risk of breast cancer following low-dose radiation exposure, *Radiology,* 131, 589, 1979.
66. Surveillance, epidemiology and end results, incidence and mortality data: 1973—77, *Natl. Cancer Inst. Monogr.,* 57, 1981.
67. Selikoff, I. J., Hammond, E. C., and Seidman, H., Latency of asbestos disease among insulation workers in the United States and Canada, *Cancer,* 46, 2736, 1979.
68. Berry, G., Newhouse, M. L., and Antonis, P., Combined effect of asbestos and smoking on mortality from lung cancer and mesothelioma in factory workers, *Br. J. Ind. Med.,* 42, 12, 1985.
69. Saracci, R., Asbestos and lung cancer: an analysis of the epidemiological evidence on the asbestos-smoking interaction, *Int. J. Cancer,* 20, 323, 1977.
70. Selikoff, I. J. and Hammond, E. C., Asbestos and smoking, *JAMA,* 242, 458, 1979.
71. Thomas, D. C., Statistical methods for analyzing effects of temporal patterns of exposure on cancer risks, *Scand. J. Work Environ. Health,* 9, 353, 1983.
72. Liddell, F. D. K., McDonald, J. C., and Thomas, D. C., Methods of cohort analysis: appraisal by application to asbestos mining, *J. R. Stat. Soc. A,* 140, 469, 1977.
73. Seidman, H., Selikoff, I. J., and Hammond, E. C., Short-term asbestos work exposure and long-term observation, *Ann. N.Y. Acad. Sci.,* 330, 61, 1979.
74. Walker, A. M., Declining relative risks for lung cancer after cessation of asbestos exposure, *J. Occup. Med.,* 26, 422, 1984.
75. Mowe, G., Gylseth, B., Hartveit, F., and Skaug, V., Fiber concentration in lung tissue of patients with malignant mesothelioma. A case-control study, *Cancer,* 56, 1089, 1985.
76. Doll, R., Morgan, L. G., and Speizer, F. E., Cancers of the lung and nasal sinuses in nickel workers, *Br. J. Cancer,* 24, 623, 1970.
77. Doll, R., Mathews, J. D., and Morgan, L. G., Cancers of the lung and nasal sinuses in nickel workers: a reassessment of the period of risk, *Br. J. Ind. Med.,* 34, 102, 1977.
78. Magnus, K., Andersen, A., and Hogetveit, A. C., Cancer of respiratory organs among workers at a nickel refinery in Norway. Second report, *Int. J. Cancer,* 30, 681, 1982.
79. Clayton, D. C. and Kaldor, J. M., Heterogeneity models as an alternative to proportional hazards in cohort studies, *Bull. Int. Stat. Inst.,* 3.2, 1, 1985.
80. *IARC Monographs on the Evaluation of the Carcinogenic Risk of Chemicals to Humans,* Vol. 23, International Agency for Research on Cancer, Lyon, 1980.
81. Lee, A. M. and Fraumeni, J. F., Jr., Arsenic and respiratory cancer in man: an occupational study, *J. Natl. Cancer Inst.,* 42, 1045, 1969.
82. Pershagen, G., Wall, S., Taube, A., and Linnman, L., On the interaction between occupational arsenic exposure and smoking and its relationship to lung cancer, *Scand. J. Work Environ. Health,* 7, 302, 1981.
83. *IARC Monographs on the Evaluation of the Carcinogenic Risk of Chemicals to Humans,* Vol. 1, International Agency for Research on Cancer, Lyon, 1972.
84. *IARC Monographs on the Evaluation of the Carcinogenic Risk of Chemicals to Humans,* Vol. 4, International Agency for Research on Cancer, Lyon, 1974.
85. *IARC Monographs on the Evaluation of the Carcinogenic Risk of Chemicals to Humans,* Vol. 29, International Agency for Research on Cancer, Lyon, 1982.
86. Goldwater, L. J., Rosso, A. J., and Kleinfeld, M., Bladder tumors in a coal tar dye plant, *Arch. Environ. Health,* 11, 814, 1965.
87. Zavon, M. R., Hoegg, U., and Bingham, E., Benzidine exposure as a cause of bladder tumors, *Arch. Environ. Health,* 27, 1, 1973.
88. Decarli, A., Peto, J., Piolatto, G., and La Vecchia, C., Bladder cancer mortality of workers exposed to aromatic amines: analysis of models of carcinogenesis, *Br. J. Cancer,* 51, 707, 1985.

89. Hoover, R. and Cole, P., Temporal aspects of occupational bladder carcinogenesis, *N. Engl. J. Med.*, 288, 1041, 1973.
90. *IARC Monographs on the Evaluation of the Carcinogenic Risk of Chemicals to Humans*, Vol. 19, International Agency for Research on Cancer, Lyon, 1979.
91. Pasternack, B. S. and Shore, R. E., Lung cancer following exposure to chloromethyl ethers, *Crit. Curr. Issues Environ. Health Hazards*, 76, 1981.
92. Armstrong, B., Endocrine factors in human carcinogenesis, in *IARC Scientific Publications No. 39, Host Factors in Human Carcinogenesis*, Bartsch, H. and Armstrong, B., Eds., International Agency for Research on Cancer, Lyon, 1982, 193.
93. *IARC Monographs on the Evaluation of the Carcinogenic Risk of Chemicals to Humans*, Vol. 21, International Agency for Research on Cancer, Lyon, 1979.
94. de Waard, F., Uterine corpus, in *Cancer Epidemiology and Prevention*, Schottenfeld, D. and Fraumeni, J. F., Jr., Eds., W. B. Saunders, Philadelphia, 1982, 901.
95. Petrakis, N. L., Ernster, V. L., and King, M.-C., Breast, in *Cancer Epidemiology and Prevention*, Schottenfeld, D. and Fraumeni, J. F., Jr., Eds., W. B. Saunders, Philadelphia, 1982, 855.
96. Henderson, B. E., Ross, R. K., Pike, M. C., and Casagrande, J. T., Endogenous hormones as a major factor in human cancer, *Cancer Res.*, 42, 3232, 1982.
97. Trichopoulos, D., MacMahon, B., and Cole, P., Menopause and breast cancer risk, *J. Natl. Cancer Inst.*, 48, 605, 1972.
98. Day, N. E. and Boice, J. D., Jr., Eds., *IARC Scientific Publications No. 52, Second Cancer in Relation to Radiation Treatment for Cervical Cancer. A Cancer Registry Collaboration*, International Agency for Research on Cancer, Lyon, 1983.
99. Brinton, L. A., Hoover, R. N., Szklo, M., and Fraumeni, J. F., Jr., Menopausal estrogen use and risk of breast cancer, *Cancer*, 47, 2517, 1981.
100. Hoover, R., Glass, A., Finkle, W. D., Azevedo, D., and Milne, K., Conjugated estrogens and breast cancer risk in women, *J. Natl. Cancer Inst.*, 67, 815, 1981.
101. Jick, H., Walker, A. M., Watkins, R. N., D'Ewart, D. C., Hunter, J. R., Danford, A., Madsen, S., Dinan, B. J., and Rothman, K. J., Replacement estrogens and breast cancer, *Am. J. Epidemiol.*, 112, 586, 1980.
102. Jick, H., Watkins, R. N., Hunter, J. R., Dinan, B. J., Madsen, S., Rothman, K. J., and Walker, A. M., Replacement estrogens and endometrial cancer, *N. Engl. J. Med.*, 300, 218, 1979.
103. Shapiro, S., Kelly, J. P., Rosenberg, L., Kaufman, D. W., Helmrich, S. O., Rosenshein, N. B., Lewis, J. L., Jr., Knapp, R. C., Stolley, P. D., and Schottenfeld, D., Risk of localized and widespread endometrial cancer in relation to recent and discontinued use of conjugated estrogens, *N. Engl. J. Med.*, 313, 969, 1985.
104. Knudson, A. G., Hereditary cancer, oncogenes, and antioncogenes, *Cancer Res.*, 45, 1437, 1985.
105. Bussey, H. J. R., *Familial Polyposis Coli*, Johns Hopkins University Press, Baltimore, 1975.
106. Greene, M. H., Clark, W. H., Tucker, M. A., Elder, D. E., Kraemer, K. H., Guerry, D. P., Witmer, W. K., Thompson, J., Matozzo, I., and Fraser, M. C., Acquired precursors of cutaneous malignant melanoma. The familial dysplastic nevus syndrome, *N. Engl. J. Med.*, 312, 91, 1985.
107. Knudson, A. G., Mutation and cancer: statistical study of retinoblastoma, *Proc. Natl. Acad. Sci. U.S.A.*, 68, 820, 1971.
108. Cavanee, W. K., Dryja, T. P., Phillips, R. A., Benedict, W. F., Godbout, R., Gallie, B. L., Murphree, A. L., Strong, L. C., and White, R. L., Expression of recessive alleles by chromosomal mechanisms in retinoblastoma, *Nature (London)*, 305, 779, 1983.
109. Benedict, W. F., Murphree, A. L., Banerjee, A., Spina, C. A., Sparkes, M. C., and Sparkes, R. S., Patient with 13 chromosome deletion: evidence that the retinoblastoma gene is a recessive cancer gene, *Science*, 219, 973, 1983.
110. Koufos, A., Hansen, M. F., Lampkin, B. C., Workman, M. L., Copeland, N. G., Jenkins, N. A., and Cavenee, W. K., Loss of alleles at loci on human chromosome 11 during genesis of Wilms' tumour, *Nature (London)*, 309, 170, 1984.
111. Orkin, S. H., Goldman, D. S., and Sallan, S. E, Development of homozygosity for chromosome 11p markers in Wilms' tumour, *Nature (London)*, 309, 172, 1984.
112. Reeve, A. E., Housiaux, P. J., Gardner, R. J. M., Chewings, W. E., Grindley, R. M., and Millow, L. J., Loss of a Harvey *ras* allele in sporadic Wilms' tumour, *Nature (London)*, 309, 174, 1984.
113. de-The, G., Geser, A,. Day, N. E., Tukei, P. M., Williams, E. H., Beri, D. P., Smith, P. G., Dean, A. G., Bornkamm, G. W., Feorino, P., and Henle, W., Epidemiological evidence for causal relationship between Epstein-Barr virus and Burkitt's lymphoma from Ugandan prospective study, *Nature (London)*, 274, 756, 1978.

114. de-The, G., Epstein-Barr virus and Burkitt's lymphoma worldwide: the causal relationship revisited, in *IARC Scientific Publications No. 60, Burkitt's Lymphoma: A Human Cancer Model*, Lenoir, G., O'Conor, G., and Olweny, C. L. M., Eds., International Agency for Research on Cancer, Lyon, 1985, 165.
115. Taub, R., Kirsch, L., Morton, C., Lenoir, C., Swan, D., Tronick, S., Aaronson, S., and Leder, P., Translocation of the *c-myc* gene into the immunoglobulin heavy chain locus in human Burkitt lymphoma and mouse plasmacytoma cells, *Proc. Natl. Acad. Sci. U.S.A.*, 79, 7837, 1982.
116. Fialkow, P. J., Martin, P. J., Nayfeld, V., Penfold, G. K., Jacobson, R. J., and Hansen, J. A., Evidence for a multistep pathogenesis of chronic myelogenous leukemia, *Blood*, 58, 158, 1981.
117. Cooper, G. M., Cellular transforming genes, *Science*, 217, 801, 1982.
118. Parkin, D. M., Hodgson, P., and Clayden, A. D., Incidence and prevalence of preclinical carcinoma of cervix in a British population, *Br. J. Obstet. Gynaecol.*, 89, 564, 1982.
119. zur Hausen, H., Herpes simplex virus in human genital cancer, *Int. Rev. Exp. Pathol.*, 25, 307, 1983.
120. Schneider, A., Kraus, H., Schuhmann, R., and Gissmann, L., Papillomavirus infection of the lower genital tract: detection of viral DNA in gynecological swabs, *Int. J. Cancer*, 35, 443, 1985.
121. Burnet, F. M., Somatic mutation and chronic disease, *Br. Med. J.*, 1, 338, 1965.
122. Hesterberg, T. W. and Barrett, J. C., Dependence of asbestos- and mineral dust-induced transformation of mammalian cells in culture on fiber dimension, *Cancer Res.*, 44, 2170, 1984.
123. Hesterberg, T. W. and Barrett, J. C., Induction by asbestos fibers of anaphase abnormalities: mechanism for aneuploidy induction and possibly carcinogenesis,. *Carcinogenesis*, 6, 473, 1985.
124. Easton, D., personal communication.

Chapter 11

TUMOR PROMOTION AND PROGRESSION IN MOUSE SKIN

Henry Hennings

TABLE OF CONTENTS

I.	Epidermal Carcinogenesis in Mouse Skin	60
II.	Two-Stage Promotion	60
III.	Mechanisms of Selection for Initiated Cells by Tumor Promoters	65
IV.	Malignant Conversion: Progression from Benign to Malignant Tumors	65
V.	Metastasis of Squamous Cell Carcinomas	69
VI.	Conclusions and Summary	69
Acknowledgments		70
References		70

I. EPIDERMAL CARCINOGENESIS IN MOUSE SKIN

The study of skin tumors is facilitated by their external location, which allows tumor development, regression, or progression to be followed without sacrificing the experimental animals. Skin tumors can be induced on the backs of mice by several different experimental protocols. The repeated topical application of a carcinogen, i.e., 7,12-dimethylbenz[a]anthracene (DMBA), produces multiple benign papillomas followed by a high incidence of squamous cell carcinomas.[1] A single large dose of the carcinogen also produces both benign and malignant tumors after a longer latent period.[2]

The initiation-promotion model of epidermal carcinogenesis has provided the basis for many conceptual advances in the understanding of the biology of carcinogenesis.[3-5] Several operational stages have been demonstrated in epidermal carcinogenesis (Table 1). Initiation by a single application of a low dose of carcinogen in Stage I causes a permanent alteration in some epidermal cells which are termed "initiated". In the absence of promoter treatment, these initiated cells do not develop into tumors, but the carcinogen-induced change is heritable. When initiated skin is exposed repeatedly to a promoting agent, i.e., 12-O-tetradecanoylphorbol-13-acetate (TPA), multiple benign papillomas develop, even when promotion is delayed for several months after initiation. The effects of individual promoter treatments are reversible. That is, papillomas do not develop after insufficient exposure of initiated skin to promoters or if the interval between treatments is prolonged. The stage of promotion, with papillomas as an end point, has been further divided into two stages which will be discussed in detail. The progression of papillomas to squamous cell carcinomas has been termed "malignant conversion".[6] This stage may proceed either spontaneously or as a result of exposure of papilloma-bearing mice to a tumor initiator. Continued treatment of papilloma-bearing mice with TPA does not affect malignant conversion. Carcinomas generally kill mice within 4 to 6 weeks, but metastasis to both the lymph nodes and lung can be demonstrated.

The irreversibility of both initiation and malignant conversion suggests a genetic mechanism for these stages. On the other hand, the reversibility of promotion indicates an epigenetic mechanism. The selection for the growth of the cells with altered developmental potential (the initiated cells) during tumor promotion may involve changes in gene expression, differentiation, and proliferation. A selective induction of terminal differentiation of normal cells by the promoter or decreased promoter toxicity to initiated cells could be involved. The high rate of proliferation in promoter-treated epidermis and in papillomas may facilitate the subsequent irreversible changes necessary for the progresson to malignancy.

II. TWO-STAGE PROMOTION

The two-stage promotion protocol developed by Slaga and co-workers[7] is shown in Table 2. In SENCAR mice, an outbred stock which was bred for sensitivity to initiation by DMBA and promotion by TPA,[8] a high yield of papillomas results from DMBA initiation and TPA promotion. Mezerein, a diterpene structurally related to TPA, is a weak promoter when treatment is begun soon after initiation. However, if initiated mice are treated once or twice with TPA before starting mezerein, a high papilloma incidence is found. This TPA treatment alone does not cause papillomas. In this protocol, Stage 1 is accomplished by TPA 1 week after initiation and Stage 2 by mezerein begun 1 week later.

Selection for the growth of initiated cells by exposure to a Stage 1 promoter could result from a direct effect of the promoter on initiated cells.[9] According to this hypoth-

Table 1
MULTISTAGE EPIDERMAL CARCINOGENESIS

Stage		Endpoint
I	Initiation	Initiated cells
II	Promotion (Stages 1 and 2)	Papillomas
III	Malignant conversion	Squamous cell carcinomas
IV	Metastasis	Metastatic squamous cell carcinomas

Table 2
TWO-STAGE PROMOTION

| Initiation (time 0) | Stage of promotion | | Papilloma incidence |
	Stage 1 (week 1)	Stage 2 (week 2 →)	
DMBA	TPA	TPA	High
DMBA	TPA	—	None
DMBA	—	Mezerein	Low
DMBA	TPA	Mezerein	High

Table 3
EXPRESSION OF PAPILLOMA PHENOTYPE

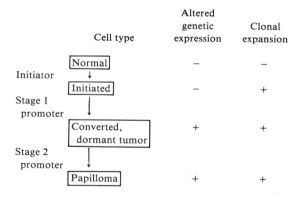

esis, treatment with a Stage 1 promoter alters the genetic expression in initiated cells to produce cells with the papilloma phenotype (Table 3). Since papillomas have a higher proliferation rate than normal epidermis, the "papilloma phenotype" implies that the cells express an inherent growth advantage over normal cells. Stage 2 promotion accomplishes the clonal expansion of cells with the papilloma phenotype, eventually resulting in papillomas. According to this altered genetic expression hypothesis, the phenotype of initiated cells is not different from normal with regard to growth.

While this hypothesis is attractive, there are other alternatives. The putative alteration in genetic expression of initiated cells cannot be demonstrated directly since initiated cells are indistinguishable from normal cells. Is this alteration in genetic expression necessary to explain two-stage promotion? Do initiated cells actually differ from papilloma cells in their phenotype or could initiation alone produce the papilloma phenotype? If initiated cells express the papilloma phenotype, then clones of cells with

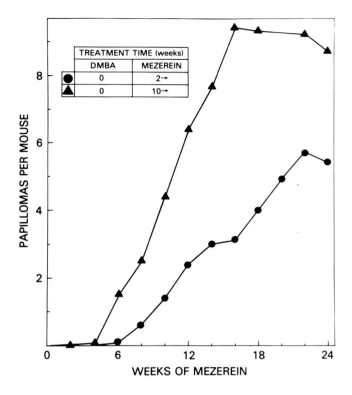

FIGURE 1. Effect on papilloma latent period of delaying promotion with mezerein. Two groups of 25 SENCAR mice (NCI-DCT Animal Program) were initiated at time 0 by a single topical treatment with 20 μg DMBA per 0.2 ml acetone. Twice weekly treatments with mezerein (4 μg/0.2 ml acetone) were begun at either 2 weeks (●) or 10 weeks (▲) after initiation. Papillomas per mouse are plotted vs. weeks of mezerein treatment.

this phenotype should expand even in the absence of promoter treatment. If this were the case, delaying promotion should shorten the latent period for papilloma development.

A survey of the skin carcinogenesis literature revealed that when promotion was delayed for 1 month or more after initiation, the time required for papilloma formation was shortened by about 3 weeks.[10] This result can be interpreted to indicate that the size of the initiated cell clones increased during the time without treatment. When promoter treatments were begun after the delay, fewer cell divisions were required for papillomas to become visible. If Stage 2 promoters act by increasing the proliferation rate of initiated cells, these agents should also be active after a delay. Thus, we designed an experiment to determine the effect on the papilloma latent period of delaying treatment with the Stage 2 promoter mezerein. In this experiment, initiation with DMBA was followed by mezerein as the promoter begun at week 2 or 10. When mezerein treatment was delayed until week 10, papillomas appeared 2 to 3 weeks earlier, and the maximum papilloma incidence was increased from 5.5 to over 9 papillomas per mouse (Figure 1). Carcinomas appeared 6 to 8 weeks earlier when mezerein was begun at week 10 (data not shown). Thus, mezerein alone becomes an effective promoter when there is a 10-week delay between initiation and the first mezerein application. Interpretation of this result in light of the two-stage promotion model suggests that Stage 1 of promotion can be accomplished spontaneously with time.

A clue which may be relevant to the action of mezerein is provided by the result that mezerein inhibits complete promotion when given simultaneously with TPA. Application of mezerein along with TPA reduced the papilloma incidence by 78% compared to TPA alone.[11] Therefore, mezerein may not be lacking some activity that complete promoters display, as suggested by Slaga.[12] Instead, mezerein may produce an inhibitory activity which prevents its potential action as a complete promoter. The cytotoxic effects of mezerein in mouse skin have been well-documented. Mufson et al.[11] found an almost complete elimination of the promotion response to mezerein when the frequency of treatments of SENCAR mice with 8.5 nmol (5.6 μg) was increased from twice to three times per week. In NMRI mice[13] and Charles River CD-1 mice,[14] toxicity from repeated mezerein treatment has been reported along with hyperplasia. In addition, Argyris[14] found that when 17 weeks of mezerein treatment preceded TPA promotion of initiated mice, the papilloma response to TPA was greatly reduced, indicating toxicity of mezerein to initiated cells. This toxicity could explain its low activity as a complete promoter.

Incomplete promoters, i.e., mezerein, may have differential promoting or antipromoting action depending on the biological state of the epidermis. The toxic effects of TPA are much greater for the first application than for subsequent TPA treatments.[15] If the relative toxic effects of mezerein are similarly decreased in hyperplastic epidermis, then the promoting effects of mezerein when applied after TPA treatment in Stage 1 may override the toxic effects. Mezerein may be less toxic to a large clone of initiated cells than to a few initiated cells soon after initiation.

Since the latent period for papilloma development is markedly shortened when Stage 2 promotion is delayed, initiated cells apparently express the papilloma phenotype and only require clonal expansion for papilloma development. Therefore, a clonal expansion hypothesis is sufficient to explain tumor promotion. As illustrated in Table 3, the critical difference between the altered genetic expression hypothesis discussed earlier and the clonal expansion hypothesis is the expression of the papilloma phenotype by initiated cells in the clonal expansion hypothesis. A proposed mechanism by which promoters may accomplish this selection for growth of initiated cells will be discussed subsequently in Section III.

Furstenberger and Marks[16] reported that Stage 1 of promotion was irreversible. They showed that after initiation and Stage 1 promotion, a delay of 10 weeks in starting Stage 2 promotion did not decrease the papilloma yield. Since the effectiveness of a Stage 2 promoter increases with time after initiation (Figure 1), the possible reversibility of Stage 1 promotion was reexamined. Papillomas were induced in a group of mice initiated with DMBA, treated with TPA at week 1 (Stage 1), then with mezerein beginning at week 10 (Stage 2). The papilloma response was identical to that found in the mice promoted only with mezerein beginning at week 10 (Figure 2). Thus, the Stage 1 effect of TPA when mezerein was given 1 week later was not seen when mezerein was delayed until week 10 (Figure 2), demonstrating the complete reversibility of Stage 1 promotion within 10 weeks. This result supports the clonal expansion hypothesis of promotion of papillomas and emphasizes the reversibility of all stages of promotion.

The relevance of the two-stage promotion protocol to the formation of malignant tumors remains to be determined since carcinoma incidences have not been reported in such experiments. The yields of papillomas and carcinomas were determined in groups of DMBA-initiated SENCAR mice promoted as follows: (1) TPA alone, (2) mezerein alone, or (3) TPA (twice in week 1) followed by mezerein. DMBA-initiated SENCAR mice promoted for 18 weeks by TPA developed 19 papillomas per mouse. Total papillomas per group of 25 mice are shown in Table 4. Promotion with mezerein alone produced 3.2 papillomas per mouse. Stage 1 TPA followed by Stage 2 mezerein induced over 14 papillomas per mouse. Surprisingly, the incidence of carcinomas in these

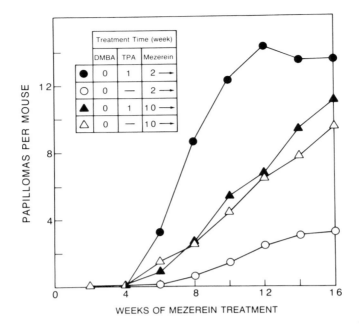

FIGURE 2. Two-stage tumor promotion in SENCAR mouse skin. Groups of 25 female SENCAR mice were treated according to the schedule as shown in the figure. Initiation was with DMBA (20 µg/0.2 ml acetone) at time 0; Stage 1 promotion was with TPA (2 µg/0.2 ml acetone) given at days 7 and 10 (week 1); and Stage 2 promotion was with mezerein (4 µg/0.2 ml) given twice per week beginning at either week 2 or 10.

Table 4
TWO-STAGE PROMOTION IN SENCAR MICE

Initiation	Stage 1	Stage 2	Total papillomas (week 18)	Number of carcinomas (week 44)
DMBA	TPA	Mezerein	321	13
DMBA	—	Mezerein	79	15
DMBA	—	TPA	457	16

Note: Three groups of 25 SENCAR mice were initiated with a single application of DMBA (20 µg/0.2 ml acetone) at time 0. One group was treated with TPA (2 µg/0.2 ml acetone) in Stage 1, at days 7 and 10 after initiation. This group was then treated with mezerein (4 µg/0.2 ml acetone, twice per week) in Stage 2 from week 2 to 31. A second group, untreated in Stage 1, received identical mezerein exposure in Stage 2. A third group, untreated in Stage 1, was treated with TPA (2 µg/0.2 ml acetone, once per week) in Stage 2 from week 2 to to 31.

three groups of mice was identical, although the papilloma yields vary nearly sixfold (Table 4). The papillomas resulting from promotion by mezerein alone are much more likely to progress to carcinomas than those promoted by TPA followed by mezerein or TPA alone.

III. MECHANISMS OF SELECTION FOR INITIATED CELLS BY TUMOR PROMOTERS

A large literature has evolved regarding the induction or inhibition of differentiation by phorbol ester treatment of a variety of cell types.[17,18] We have demonstrated in epidermal cell culture that there are at least two populations of basal cells, one which is induced by TPA to terminally differentiate while the other continues proliferation.[19] Reiners and Slaga[20] have reported similar findings after TPA treatment in vivo, and Parkinson et al.[21] have described a subpopulation of human keratinocytes which is resistant to induction of terminal differentiation by TPA. Promoters could act by allowing cells which would normally cease proliferation and terminally differentiate upon leaving the basal layer to continue proliferation in the suprabasal layers. The proliferation of cells away from the basement membrane in papillomas[22] supports this hypothesis. Raick[23] demonstrated the TPA-induced appearance of "dark cells", poorly-differentiated, ribosome-rich cells which were first seen in embryonic development of the epidermis. More recently, Slaga[12] has emphasized the importance of these cells in the early stages of promotion. While initiated cells appear to lack the ability to respond to the differentiative stimulus of tumor promoting phorbol esters,[24] other hypotheses for the promoter-induced selection for growth of initiated cells must also be considered. Nonphorbol ester tumor promoters may accomplish promotion by other selective mechanisms. Promoter effects which could be relevant include increased proliferation,[25] altered gene expression,[9] selective toxicity, immunological effects, inhibition of cellular communication,[26] altered responsiveness to growth factors, induction of ornithine decarboxylase,[27] and increased prostaglandin synthesis.[28]

IV. MALIGNANT CONVERSION: PROGRESSION FROM BENIGN TO MALIGNANT TUMORS

In initiation-promotion experiments in various strains of Swiss mice, most squamous cell carcinomas progress from papillomas, but the rate of conversion is low.[29] The factors necessary for this malignant conversion have not been defined. After DMBA initiation (Stage I) and limited TPA promotion (Stage II) in SENCAR mice, continued TPA treatment of these papilloma-bearing mice was compared to repeated exposure to an initiator for the effect on carcinoma development (Stage III).[6] Continued TPA treatment in Stage III does not increase carcinoma development relative to treatment with the solvent acetone (Figure 3), but the treatment of papilloma-bearing mice with genotoxic tumor initiators markedly accelerates and enhances malignant conversion. Stage III treatments with N-methyl-N'-nitro-N-nitrosoguanidine (MNNG), 4-nitroquinoline-N-oxide (4-NQO), and urethane for 30 weeks were all effective when given after DMBA in Stage I and TPA in Stage II (Figure 3). Treatment of an uninitiated control group with TPA in Stage II and MNNG for 30 weeks in Stage III produced 3.0 papillomas per mouse and a 14% carcinoma yield.[6] In a second experiment (data not shown), elimination of either DMBA in Stage I or TPA treatment in Stage II before MNNG treatment in Stage III did not reduce the incidence of papillomas or carcinomas. Thus, the activity of MNNG as a complete carcinogen or cocarcinogen limits its usefulness as a Stage III agent. In subsequent three-stage experiments, the low complete carcinogenic and cocarcinogenic activity of both 4-NQO and urethane (Table 5) has verified their utility as Stage III agents.

In order to demonstrate the three-stage carcinogenesis protocol in mice other than SENCARs, a similar experiment was performed in Charles River CD-1 mice. In CD-1 mice initiated with DMBA (Stage I) and promoted with TPA for 12 weeks (Stage II), many papillomas regressed after TPA exposure was terminated (Figure 4). When TPA

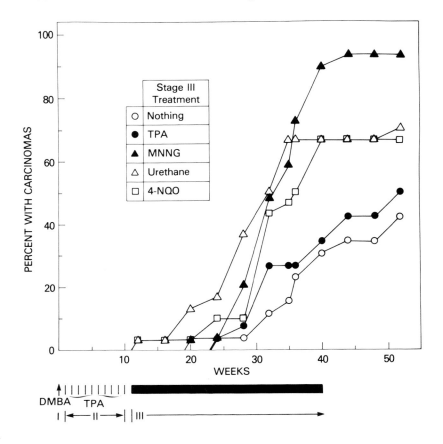

FIGURE 3. A comparison of carcinoma development in papilloma-bearing mice treated in Stage III with acetone, TPA, MNNG, urethane, or 4-NQO. Tumors were induced in groups of 30 7- to 8-week-old female SENCAR mice by the treatment schedule indicated at the bottom of the figure. Stage I initiation by a single topical application of DMBA (20 µg/0.2 ml acetone) was followed by Stage II promotion with TPA (2.5 µg/0.2 ml acetone weekly for 10 weeks) in all groups. Weekly Stage III treatments during weeks 11 to 40 were as follows: topical application of 0.2 ml acetone (O); 0.2 ml acetone containing 2.5 µg TPA (●), 120 µg MNNG (▲), or 250 µg 4-NQO (□); or i.p. injection of 20 mg urethane per 0.2 ml saline (△). Papillomas and suspected carcinomas were counted weekly. The "percent with carcinomas", calculated as the number of carcinoma-bearing mice divided by the number of mice alive in each group when the first carcinoma appeared and expressed as a percentage, is plotted against weeks of the experiment.

was continued, the papilloma incidence increased until week 28. However, the final carcinoma yield was virtually the same in groups receiving continuous TPA or just acetone solvent in Stage III (Figure 4). Since the number of papillomas remaining at the time carcinomas were first seen was markedly reduced in the acetone group, TPA-dependent papillomas are at very low risk for carcinoma formation. With urethane or 4-NQO in Stage III, papillomas regressed at a rate similar to that seen in acetone-treated mice (Figure 4). In addition, the maximum papilloma incidence at 16 weeks was reduced 30% by 4-NQO. Either 4-NQO or urethane in Stage III clearly increased the rate of carcinoma formation and the final carcinoma yield relative to TPA or acetone treatment (Figure 4). The low tumor yield in control groups not exposed to TPA in Stage II demonstrates the requirement for a papilloma stage as a precursor to carcinomas. Stage I initiation followed by solvent treatment in Stage II and urethane

Table 5
IMPORTANCE OF TPA IN STAGE II FOR INDUCTION OF CARCINOMAS BY STAGE III AGENTS

Treatment in stage		SENCAR		CD-1	
II	III	Papillomas per mouse	Percent with carcinomas	Papillomas per mouse	Percent with carcinomas
TPA	4-NQO	29.0	72	9.4	55
—	4-NQO	2.8	21	0.7	8
TPA	Urethane	35.0	60	13.3	64
—	Urethane	1.1	7	0.03	0

Note: Stage I initiation in all four groups was by a single treatment with 20 μg DMBA in SENCAR mice and 50 μg DMBA in CD-1 mice.

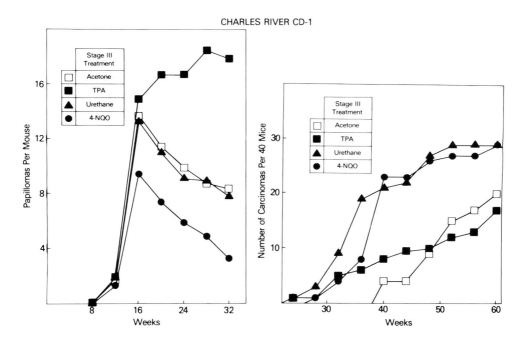

FIGURE 4. Papilloma and carcinoma incidence in a three-stage carcinogenesis experiment in CD-1 mice. Groups of 40 female CD-1 mice (Charles River, Kingston, N.Y.) were initiated with 50 μg DMBA per 0.2 ml acetone (Stage I) and promoted with 12 weekly applications of 10 μg TPA per 0.2 ml acetone (Stage II). The indicated Stage III treatments once weekly from week 13 to 52 were as follows: acetone, 0.2 ml (□); TPA, 10 μg/0.2 ml acetone (■); urethane, 20 mg i.p. (▲); 4-NQO, 250 μg/0.2 ml acetone (•).

or 4-NQO in Stage III produces few papillomas (3 with urethane and 31 with 4-NQO per 40-mouse group) and almost no malignancies (0 with urethane and 4 with 4-NQO) (Table 5). Comparative results with the more sensitive SENCAR mice are also shown in this table.

In the experiment shown in Figure 4, the percent conversion of papillomas to carcinomas was calculated at week 60 based on the maximum number of papillomas which developed in each group. The percent malignant conversion was 2.5% with continued TPA treatment compared to 3.7% in the acetone control group. Burns et al.[29] have

68   Mechanisms of Environmental Carcinogenesis

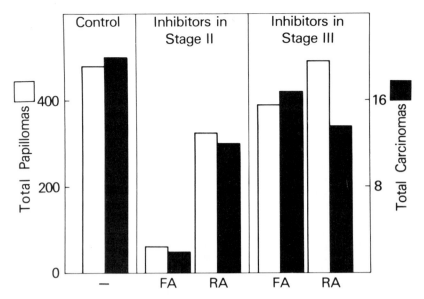

FIGURE 5. Effects of FA and RA on papilloma and carcinoma development in a three-stage carcinogenesis experiment. Groups of 40 Charles River CD-1 mice were initiated with 50 μg DMBA (Stage I) and promoted with 10 μg TPA once per week for 12 weeks (Stage II). Stage III i.p. injections of 20 mg urethane were given beginning at week 13. The inhibitors FA (1 μg/0.2 mℓ acetone) and RA (5.1 μg/10.2 mℓ acetone) were applied topically 30 min before each weekly TPA treatment in Stage II or 30 min before each urethane injection in Stage III. The cumulative papilloma and carcinoma incidences are shown at week 38.

reported a similar decrease with continued TPA treatment. The final percent conversion was increased to 5.5% with urethane and to 7.9% with 4-NQO. The difference between mice treated with acetone or TPA and mice treated with genotoxic agents in Stage III was much more striking at earlier times. At 40 weeks, the percent conversion was 0.75% with acetone, 1.2% with TPA, 4.2% with urethane, and 6.3% with 4-NQO. Thus, genotoxic agents in Stage III not only increase the number of carcinomas, but dramatically accelerate their rate of appearance.

Two known inhibitors of TPA promotion, fluocinolone acetonide (FA) and retinoic acid (RA), were tested for their effects on papilloma and carcinoma incidence when given to CD-1 mice in either Stage II or III. After DMBA initiation, treatment with FA 30 min prior to each TPA treatment in Stage II inhibited papilloma development by almost 90% (Figure 5). When these mice were then treated with urethane in Stage III, a similar decrease in carcinoma yield was seen. RA in Stage II was less effective (many papillomas developed after week 12), but the reduction in number of papillomas corresponded well with the reduction in carcinomas. When FA and RA were given to papilloma-bearing mice in Stage III, carcinoma development was not inhibited. Thus, inhibitors effective in Stage II are not necessarily effective in Stage III. These results suggest the necessity of a papilloma stage in the development of carcinomas, at least in initiation-promotion or initiation-promotion-conversion experiments. This question was approached directly by mapping the location and approximate size of all papillomas and carcinomas in a group of 30 SENCAR mice initiated with DMBA and promoted for 5 weeks with TPA. This limited promotion protocol produces few papillomas, but those which appear have a high risk of conversion to malignancy.[35] Individual papillomas were followed to see whether they persisted, regressed, or progressed to carcinomas. The 25 carcinomas seen in these mice all appeared to develop from papillomas.

Table 6
METASTASES FROM SKIN SQUAMOUS CELL CARCINOMAS

Stage III treatment	Number of mice with skin squamous cell carcinomas	Number of mice with metastatic squamous cell carcinomas	Percent metastasis	Metastatic site	
				Lymph node	Lung
Acetone	18	3	17	3	2
TPA	13	3	23	3	0
MNNG	20	6	30	5	1
Urethane	24	8	33	3	6
4-NQO	20	3	15	3	1

Note: Groups of 40 female CD-1 mice were initiated with 50 µg DMBA (Stage I) and promoted with 12 weekly applications of 10 µg TPA (Stage II). The indicated Stage III treatments once weekly from week 13 to 52 were as follows: acetone, 0.2 mℓ; TPA, 10 µg; MNNG, 120 µg; urethane, 20 mg i.p.; 4-NQO, 250 µg. Mice were sacrificed at week 60. Some animals developed metastases in both lymph node and lung as indicated.

V. METASTASIS OF SQUAMOUS CELL CARCINOMAS

In the three-stage carcinogenesis experiment described in Figure 4, complete autopsies were performed on all animals and possible metastatic lesions were examined histologically. After DMBA initiation (12 weeks of TPA promotion and either continued TPA or acetone treatment for a further 40 weeks) 17 to 23% of the mice with one or more carcinomas also developed metastatic carcinomas (Table 6). With MNNG in Stage III, the metastasis rate increased to 30%, but with 4-NQO, a metastasis rate of only 15% was found. With acetone, TPA, MNNG, or 4-NQO, a total of 14 metastases were seen in lymph nodes with only 4 in the lungs. Both the pattern and frequency of metastasis were altered in the mice exposed to urethane. Metastases were seen in 33% of the carcinoma-bearing mice, with three in lymph nodes and six in lungs. The lung squamous cell carcinomas seen after Stage III urethane exposure were not primary urethane-induced lesions. Identical urethane treatment of uninitiated mice or initiated mice not treated with TPA in Stage II produced no squamous cell carcinomas of the skin or lung. Primary lung adenomas were observed in about one half the mice in these two groups as well as in the group which developed metastatic lung tumors.

VI. CONCLUSIONS AND SUMMARY

The multistage nature of experimental epidermal carcinogenesis is well-established. Cells progress through stages of initiation, promotion of benign tumors, and then conversion to malignant tumors capable of metastasis. Similar stages have been described in carcinogenesis models in other tissues.[30] Epidemiological evidence has also suggested a multistage model for tumor development in man.[31,32]

The promotion of papilloma formation has been divided into two stages operationally. The results presented here suggest that the mechanism of action of Stages 1 and 2 promoters may not differ. The greater toxicity of the Stage 2 agent mezerein to initiated cells when given soon after initiation may explain its weak promoting activity in a standard protocol. When the first mezerein treatment is delayed until 10 weeks after initiation, mezerein is an effective promoter. Thus, Stage 1 of promotion is accomplished spontaneously during this 10 week delay. An increase in clone size of initiated cells during the 10 weeks between initiation and the commencement of mezerein treatment is consistent with this result. If this interpretation of the result is correct, only clonal expansion of initiated cells is required for papilloma development. A direct effect of promoters on initiated cells to alter their program of genetic expression does not appear to be necessary in promotion. Stage 1 promotion by TPA is completely reversible when 9 weeks elapsed between Stage 1 TPA and Stage 2 mezerein exposure. Thus, all stages of promotion may be explained by an epigenetic mechanism.

Papillomas induced by initiation-promotion are necessary precursor lesions of squamous cell carcinomas. They are heterogeneous in their tendency to progress to carcinomas. The probability of malignant conversion is highest for the papillomas induced by the first few TPA treatments and lowest for TPA-dependent papillomas. Malignant conversion is not affected by continued TPA treatment, although other promoters, i.e., mezerein, benzoyl peroxide,[33] and teleocidin,[34] could differ from TPA in this respect. Genotoxic agents, i.e., 4-NQO and urethane, enhance the conversion of papillomas to carcinomas in Charles River CD-1 mice as well as SENCAR mice.

Carcinomas induced by a three-stage protocol (Stage I DMBA initiation, limited TPA promotion in Stage II, followed by treatment with a genotoxic agent or continued TPA in Stage III) vary in their potential for metastasis. The lymph node metastasis rate was similar in Charles River CD-1 mice treated with either acetone, TPA, MNNG, urethane, or 4-NQO in Stage III. However, urethane in Stage III increased the metastatic rate to the lung from <12 to 25%.

ACKNOWLEDGMENTS

These studies were planned in collaboration with Dr. Stuart H. Yuspa and were carried out under NCI Contract NO1-CP-41015 awarded to Microbiological Associates, Bethesda, Md. Thanks are due to Dr. Edwin F. Spangler, Robert Shores, Patricia Mitchell, and Deborah Devor of Microbiological Associates, to Dr. Kjell M. Elgjo and Dr. Stuart H. Yuspa for histological examination of tumors, to Dr. Stuart H. Yuspa and Dr. James E. Strickland for numerous discussions, and to Maxine Bellman and Sandy White for typing the manuscript.

REFERENCES

1. Shubik, P., The growth potentialities of induced skin tumors in mice. The effects of different methods of chemical carcinogenesis, *Cancer Res.*, 10, 713, 1950.
2. Iversen, O. H. and Iversen, U., A study of epidermal tumorigenesis in the hairless mouse with a single and with repeated applications of 3-methylcholanthrene at different dosages, *Acta Pathol. Microbiol. Scand.*, 62, 305, 1964.
3. Boutwell, R. K., The function and mechanism of promoters of carcinogenesis, *CRC Crit. Rev. Toxicol.*, 2, 419, 1974.
4. Burns, F., Albert, R., Altshuler, B., and Morris, E., Approach to risk assessment for genotoxic carcinogens based on data from the mouse skin initiation-promotion model, *Environ. Health Perspect.*, 50, 309, 1983.
5. Hennings, H., Experimental skin carcinogenesis, in *Pathophysiology of Dermatologic Diseases*, Soter, N. A. and Baden, H. P., Eds., McGraw Hill, New York, 1984, chap. 24.
6. Hennings, H., Shores, R., Wenk, M. L., Spangler, E. F., Tarone, R., and Yuspa, S. H., Malignant conversion of mouse skin tumors is increased by tumor initiators and unaffected by tumor promoters, *Nature (London)*, 304, 67, 1983.
7. Slaga, T. J., Fischer, S. M., Nelson, K., and Gleason, G. L., Studies on the mechanism of skin tumor promotion: evidence for several stages in promotion, *Proc. Natl. Acad. Sci. U.S.A.*, 77, 3659, 1980.
8. Hennings, H., Devor, D., Wenk, M. L., Slaga, T. J., Former, B., Colburn, N. H., Bowden, G. T., Elgjo, K., and Yuspa, S. H., Comparison of two-stage epidermal carcinogenesis initiated by 7,12-dimethylbenz[a]anthracene or N-methyl-N'-nitro-N-nitrosoguanidine in newborn and adult SENCAR and Balb/c mice, *Cancer Res.*, 41, 773, 1981.

9. Furstenberger, G. and Marks, F., Growth stimulation and tumor promotion in skin, *J. Invest. Dermatol.*, 81, 157s, 1983.
10. Hennings, H. and Yuspa, S. H., Two-stage tumor promotion in mouse skin: an alternative explanation, *J. Natl. Cancer Inst.*, 74, 735, 1985.
11. Mufson, R. A., Fischer, S. M., Verma, A. K., Gleason, G. L., Slaga, T. J., and Boutwell, R. K., Effects of 12-O-tetradecanoylphorbol-13-acetate and mezerein on epidermal ornithine decarboxylase activity, isoproterenol-stimulated levels of cyclic adenosine 3′:5′-monophosphate, and induction of mouse skin tumors in vivo, *Cancer Res.*, 39, 4791, 1979.
12. Slaga, T. J., Overview of tumor promotion in animals, *Environ. Health Perspect.*, 50, 3, 1983.
13. Furstenberger, G., Berry, D. L., Sorg, B., and Marks, F., Skin tumor promotion is a two-stage process, *Proc. Natl. Acad. Sci. U.S.A.*, 78, 7722, 1981.
14. Argyris, T. S., Nature of the epidermal hyperplasia produced by mezerein, a weak tumor promoter, in initiated skin of mice, *Cancer Res.*, 43, 1768, 1983.
15. Raick, A. N., Thumm, K., and Chivers, B. R., Early effects of 12-O-tetradecanoylphorbol-13-acetate on the incorporation of tritiated precursor into DNA and the thickness of the interfollicular epidermis, and their relation to tumor promotion in mouse skin, *Cancer Res.*, 32, 1562, 1972.
16. Furstenberger, G., Sorg, B., and Marks, F., Tumor promotion by phorbol esters in skin: evidence for a memory effect, *Science*, 220, 89, 1983.
17. Rovera, G., O'Brien, T. G., and Diamond, L., Induction of differentiation in human promyelocytic leukemia cells by tumor promoters, *Science*, 204, 868, 1979.
18. Lotem, J. and Sachs, L., Regulation of normal differentiation in mouse and human myeloid leukemia cells by phorbol esters and the mechanism of tumor promotion, *Proc. Natl. Acad. Sci. U.S.A.*, 76, 5158, 1979.
19. Yuspa, S. H., Ben, T., Hennings, H., and Lichti, U., Divergent responses in epidermal basal cells exposed to the tumor promoter 12-O-tetradecanoylphorbol-13-acetate, *Cancer Res.*, 42, 2344, 1982.
20. Reiners, J. J., Jr., and Slaga, T. J., Effects of tumor promoters on the rate and commitment to terminal differentiation of subpopulations of murine keratinocytes, *Cell*, 32, 247, 1983.
21. Parkinson, E. K., Grabham, P., and Emmerson, A., A subpopulation of cultured human keratinocytes which is resistant to the induction of terminal differentiation-related changes by phorbol-12-myristate-13-acetate: evidence for an increase in the resistant population following transformation, *Carcinogenesis*, 4, 857, 1983.
22. Fukuda, M., Okamura, K., Rohrbach, R., Bohm, N., and Fujita, S., Changes in cell population kinetics during epidermal carcinogenesis, *Cell Tissue Kinet.*, 11, 611, 1978.
23. Raick, A. N., Ultrastructural, histological, and biochemical alterations produced by 12-O-tetradecanoylphorbol-13-acetate on mouse epidermis and their relevance to skin tumor promotion, *Cancer Res.*, 33, 269, 1973.
24. Hennings, H., Ben, T., and Yuspa, S. H., Treatment of epidermal cell lines with 12-O-tetradecanoylphorbol-13-acetate (TPA): lack of differentiative response and variable proliferative response, *Proc. Am. Assoc. Cancer Res.*, 25, 146, 1984.
25. Argyris, T. S., Epidermal growth following a single application of 12-O-tetradecanoylphorbol-13-acetate in mice, *Am. J. Pathol.*, 98, 639, 1980.
26. Fitzgerald, J. and Murray, A., Inhibition of intercellular communication by tumor-promoting phorbol esters, *Cancer Res.*, 40, 2435, 1980.
27. O'Brien, T. G., Simsiman, R. C., and Boutwell, R. K., Induction of the polyamine-biosynthetic enzymes in mouse epidermis by tumor-promoting agents, *Cancer Res.*, 35, 1662, 1975.
28. Marks, F., Prostaglandins, cyclic nucleotides and the effect of phorbol ester tumor promoters on mouse skin *in vivo*, *Carcinogenesis*, 4, 1465, 1983.
29. Burns, F. J., Vanderlaan, M., Snyder, F., and Albert, R. E., Induction and progression kinetics of mouse skin papillomas, in *Carcinogenesis: A Comprehensive Survey*, Vol. 2, Slaga, T. J., Sivak, A., and Boutwell, R. K., Eds., Raven Press, New York, 1978, 91.
30. Slaga, T. J., Sivak, A., and Boutwell, R. K., Eds., *Carcinogenesis: A Comprehensive Survey*, Vol. 2, Raven Press, New York, 1978.
31. Moolgavkar, S. H. and Knudson, A. G., Jr., Mutation and cancer: a model for human carcinogenesis, *J. Natl. Cancer Inst.*, 66, 1037, 1981.
32. Peto, J., Early- and late-stage carcinogenesis in mouse skin and in man, in *Models, Mechanisms and Etiology of Tumor Promotion*, Borzsonyi, M., Day, N. E., Lapis, K., and Yamasaki, H., Eds., Oxford University Press, New York, 1984, 359.
33. Slaga, T. J., Klein-Szanto, A. J. P., Triplett, L. L., Yotti, L. P., and Trosko, J. E., Skin tumor-promoting activity of benzoyl peroxide, a widely used free radical generating compound, *Science*, 213, 1023, 1981.
34. Fujiki, H., Suganuma, M., Matsukura, N., Sugimura, T., and Takayama, S., Teleocidin from *Streptomyces* is a potent promoter of mouse skin carcinogenesis, *Carcinogenesis*, 3, 895, 1982.
35. Hennings, H. and Yuspa, S. H., unpublished results.

Chapter 12

CELLULAR AND MOLECULAR MECHANISMS OF MULTISTEP CARCINOGENESIS IN CELL CULTURE MODELS

J. Carl Barrett and William F. Fletcher

TABLE OF CONTENTS

I.	Introduction	74
II.	Description of Carcinogen-Induced Multistep Carcinogenesis in Cell Culture Systems	74
III.	Oncogenes and Multistep Carcinogenesis	77
IV.	Mechanisms of Early Stages in Cell Transformation	79
	A. Morphological Transformation of Syrian Hamster Embryo Cells	79
	B. Early Morphological and Growth Alterations in Epithelial Cell Systems	81
	C. Immortality	86
V.	Mechanisms of Late Stages in Neoplastic Transformation of Cells in Culture	91
	A. Evidence for Multiple Steps	91
	B. Possible Mechanisms for Late Stages of Neoplastic Development of Cells in Culture	96
	1. Oncogenes	96
	2. Tumor Suppression Genes	98
	3. Differentiation Defects and Neoplastic Transformation of Cells in Culture	99
	4. Promotion of Cell Transformation	99
	5. Single-Step Neoplastic Transformation and Other Inconsistencies with the Immortalization/Transformation Model	100
VI.	Comparison of Human and Rodent Cell Transformation	101
	A. Induction of Tumorigenicity in Human Cells	102
	B. Induction of Morphological Transformation	103
	C. Induction of Anchorage Independence	103
	D. Induction of Immortality	104
	E. Possible Differences Between Human and Rodent Cell Transformation	105
VII.	Conclusion	105
References		106

I. INTRODUCTION

There are several lines of evidence which support the hypothesis that carcinogenesis is a multistep process (Table 1). Studies of the temporal sequence of histologically distinct stages in the neoplastic development of cancers of the skin, respiratory tract, mammary glands, uterus, and prostate have provided pathological evidence that tumors develop by a progressive process through a series of qualitatively different stages.[1] A two-stage model of carcinogenesis has been clearly defined in epidermal carcinogenesis by the separate action of different classes of chemicals.[2] In addition the progression of papillomas to carcinomas represents a further step in this model (Chapter 11). Certain human cancers, i.e., retinoblastoma[4,5] and carcinoma of the colon and rectum,[6] appear to arise by multiple changes occurring at a very high rate in genetically predisposed individuals. Evidence has been presented that these diseases arise by two mutational events, one of which can be a germ line mutation and the other a somatic mutation.[5,7,8]

Mathematical analyses of the age-specific incidence of certain kinds of cancers are also consistent with a multistage model of neoplastic development. Several studies have shown that the logarithm of the incidence of various neoplasms is related linearly to the logarithm of age; the slope of this linear relationship suggests that multiple, independent changes must accumulate before neoplastic transformation is expressed. The number of required events ranges from three to seven, according to the type of neoplasm.[9-11] Some of the changes may be clonal expansions of intermediate cells in the neoplastic progression; therefore the actual number of qualitative changes may be only two or three.[10,11]

Studies using cell culture models of carcinogenesis have clearly demonstrated the progressive, multistep nature of neoplastic transformation.[12-16] The need for multiple genetic changes in a normal cell for neoplastic conversion has also been clearly shown by DNA transfection experiments of oncogenes into normal or partially altered cells.[17-22] The primary focus of this chapter is on the cellular and molecular mechanisms of multistep carcinogenesis; therefore, this discussion will center on a further analysis of the above two types of studies. Cell culture studies offer the advantages of allowing one quantitatively to study cellular, biochemical, and molecular properties of defined populations of cells by techniques not readily applicable to in vivo studies. However, the artificial environment of cells in culture, the continued stimulation of proliferation of these cells, and the lack of cellular interactions evident in a tissue are major differences between in vivo and in vitro models. When possible, comparisons of the changes observed in cell cultures will therefore be made to cellular-molecular events in cells in vivo. This extrapolation is of utmost importance for fully understanding the mechanisms of carcinogenesis, and is discussed more fully in Chapter 13.

II. DESCRIPTION OF CARCINOGEN-INDUCED MULTISTEP CARCINOGENESIS IN CELL CULTURE SYSTEMS

Berwald and Sachs[24,25] were the first to demonstrate the induction by chemical carcinogens of neoplastic transformation of mammalian cells in culture. Following exposure to 3-methylcholanthrene or benzo(a)pyrene (BaP), Syrian hamster embryo fibroblasts were observed to escape "cellular senescence", the process which limits the proliferation of normal cells in culture to a certain number of cell divisions.[26] The carcinogen-treated cells had an indefinite life span in culture, which has been termed immortalization,[27] and were morphologically transformed. The morphologically altered cell lines ultimately produced malignant fibrosarcomas when injected into syn-

Table 1
LINES OF EVIDENCE FOR MULTISTAGE MODELS OF NEOPLASTIC DEVELOPMENT

1. Pathological observations of tumors show a temporal sequence of histologically distinct stages in many neoplasms
2. Chemical carcinogenesis studies in mouse skin and other tissues reveal that carcinogenesis involves stages of initiation, promotion, and progression
3. Individuals with genetic traits manifested by an early occurrence of cancer (e.g., familial retinoblastoma, adenomatosis of the colon and rectum) may inherit one germ line mutation for cancer and develop cancer upon acquisition of a second, somatic cell mutation
4. Mathematical models based on age-specific tumor incidence curves are consistent with a multistage model of neoplastic development involving 2 to 7 independent changes
5. Cell culture studies with chemical carcinogen-induced neoplastic transformation indicate that multiple changes must occur in the transformation of a normal cell to a tumor cell
6. Studies with viral and cellular oncogenes indicate that at least two cooperating oncogenes are necessary for neoplastic conversion of normal cells

geneic hamsters,[24] whereas untreated Syrian hamster embryo cells were nontumorigenic. Tumors can be produced by injections of <10 transformed cells while 10^7 to 10^9 normal Syrian hamster embryo cells are nontumorigenic even after several passages in culture.[28-30] Studies by Berwald and Sachs,[24,25] Huberman and Sachs,[29] and DiPaolo et al.[31-33] showed that transformation of these cells could be observed and quantitated by examining the morphology of colonies that form from single cells following carcinogen exposure. Known chemical carcinogens, but not structurally related noncarcinogens, induce colonies with an altered morphology identified by loss of orderly growth, crisscrossing of cells, increased basophilia, and increased nuclear-to-cytoplasm ratios.

Morphological transformation is the earliest phenotypic marker that can be detected in Syrian hamster embryo cells following carcinogen treatment.[12] Other phenotypic changes occur following carcinogen exposure with a different temporal sequence. Cells with an enhanced fibrinolytic activity are observed at 2 to 3 weeks, whereas cells with the ability to grow when suspended in a semisolid medium (i.e., soft agar) or able to form tumors when injected into syngeneic animals are not observed until 6 to 15 weeks (30 to 100 population doublings) after carcinogen treatment.[12] In this cellular system the ability to grow in agar correlates very well with tumorigenicity for chemical carcinogen-induced transformants.[30]

The inability to detect cells which grow in agar or as tumors in animals is not due to insensitivity of the assays to detect cells with these phenotypes.[12] Rather, cells expressing the early phenotypic changes are preneoplastic and require additional changes to become tumorigenic. This hypothesis is supported by observations of a number of laboratories working with this cell system. Berwald and Sachs[24,25] initially observed a very high frequency of cell transformation, up to 25.6% morphologically transformed colonies, at 8 days after BaP treatment. However, they were unable to isolate viable clones 15 to 20 days after introduction of the carcinogen. Other laboratories have experienced similar difficulties in isolating morphologically transformed colonies. These observations are inconsistent with the notion that fully transformed cells are present soon after carcinogen treatment. After allowing further growth in culture for 40 to 50 days, Berwald and Sachs[24,25] were able to isolate and establish cell lines from 90% of the morphologically transformed clones. When they tested the cells for tumor formation after 72 days in culture, the tumors grew initially, but then regressed. Growth of the cells in vitro for 98 days markedly decreased this initial regression, suggesting a further alteration of the cell population with cultivation. Borek and Sachs[34] observed morphological transformation of hamster cells by X-irradiation, but

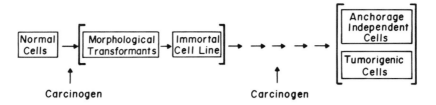

FIGURE 1. Neoplastic progression of Syrian hamster embryo cells.

the cells did not develop the ability to grow in soft agar or produce progressively growing tumors in vivo. DiPaolo and Donovan[35] also reported the inability to produce tumors from carcinogen-treated Syrian hamster embryo cells shortly after treatment. At least 74 days in culture were required before tumorigenicity was observed after injection of very large numbers (8 to 20 × 10^7) of cells. Kuroki and Sato[36] extensively studied the neoplastic transformation in vitro of Syrian hamster embryo cells by 4-nitroquinoline-1-oxide (4-NQO). The earliest time at which cells capable of producing progressively growing tumors were obtained was 49 days after the initial carcinogen treatment. In contrast, early changes in morphology were seen 3 to 4 days after treatment, and histological evidence of transformation was observed as early as 23 days after treatment. Furthermore, Kuroki and Sato identified three different stages in the neoplastic development of the transformed cultures, which they cite as evidence for progression in vitro. Similar studies with Syrian hamster embryo cells following treatment with 4-NQO were reported by Kakunaga and Kamahora.[37,38]

Studies by Barrett and Ts'o[12] quantitatively analyzed the frequency of carcinogen-treated Syrian hamster embryo cells demonstrating different phenotypic alterations. Morphologically transformed cells and cells with enhanced fibrinolytic activity occurred at high frequencies within 1 to 2 weeks after carcinogen treatment, while anchorage-independent or tumorigenic cells were not detectable at this time. Since sensitive assays for anchorage-independence and tumorigenicity were used, the inability to detect cells with these phenotypes until 6 to 15 weeks after treatment indicated that these are secondary changes not directly induced by the carcinogen. The morphologically transformed colonies and colonies with enhanced fibrinolytic activity gave rise to immortal cell lines with a high frequency (10 to 50% of the colonies). Anchorage-independent and tumorigenic cells developed within the immortal cell populations by a second, qualitative change in the cells. Selection of preexisting cells with these changes was ruled out. Newbold et al.[27] have also shown that immortal cells have a higher propensity to develop into neoplastic and anchorage-independent cells than normal cells. These results clearly demonstrate that the process of neoplastic development of Syrian hamster embryo cells is a multistep, progressive process requiring at least two heritable alterations, immortality and anchorage-independent growth, before they become tumorigenic. A diagram of the most common pathway for neoplastic transformation of these cells is shown in Figure 1.

This conclusion is further supported by the ability to isolate intermediate cells or preneoplastic cells which have acquired some, but not all, of the properties necessary to be tumorigenic.[39] These cells can be shown to be preneoplastic because, while they are nontumorigenic, they have an increased propensity relative to normal cells to become neoplastic after further growth in vitro, growth in vivo under certain conditions, or after treatment with carcinogens or defined oncogenes.[21,39,40]

Similar stages of neoplastic progression of other cell types in culture have also been may be late events, which is consistent with the effects of carcinogens and mutagens on the conversion of papillomas to carcinomas.[3]

It is interesting to compare the differentiation-altered mouse keratinocytes with the enhanced growth variants of RTE cells discussed in the preceding section. Certain similarities exist in these transformants of different cells. Both altered populations are selected on the basis of continued growth of the transformants after imposition of selective conditions which cause the normal cells to cease proliferation. In the case of the mouse keratinocytes, it has been convincingly shown that the cells stop proliferating in high calcium media due to terminal differentiation.[90] The basis for cessation of growth of the normal rat tracheal epithelial cells is less clear. The cells do enlarge, appear squamous-like, and slough from the dish.[83,88] Morphologically this process resembles terminal differentiation of epidermal cells. Also, tracheal cells can undergo keratinocyte differentiation under other conditions.[103] Therefore, these cells may stop growing due to terminal differentiation like the mouse keratinocytes which would suggest that enhanced growth variants of rat tracheal cells and differentiation-altered mouse keratinocytes are fundamentally similar. Further studies are required to test this hypothesis. Both types of transformants are induced by carcinogens, and the frequency of altered cells per surviving cell is relatively high, i.e., >1%, for each cell type.[83,88,91,104] This high frequency supports the conclusion drawn above that these changes are unlikely to arise as a result of a point mutation of the *ras* oncogenes.

One important study provides evidence for the role of the *ras* oncogene in the early stages of epithelial cell transformation. Yoakum et al.[105] have reported that transfection of human bronchial epithelial cells with the viral Ha-*ras* oncogene results in differentiation and/or growth alterations of the cells. The normal bronchial cells fail to grow and undergo terminal, squamous differentiation in medium containing blood-derived serum; in contrast, *ras*-transfected cells continue to grow under these selective conditions. These altered cells, which express the v-Ha-*ras* DNA, grow for 10 to 12 cell divisions after the control cultures differentiate and then become quiescent and senesce. When the cells are maintained at high cell density prior to senescence and crisis, foci of cells with partial loss of contact-inhibition appear, and some of these foci can be subcultured and grow indefinitely (>120 generations). These immortal cell lines retain the differentiation alteration, i.e., they are not induced to squamous terminal differentiation by different inducers; however, the cells do not grow in soft agar. When the cells are injected in nude mice, small nodules of <0.1 cm in size develop at a low frequency, but these nodules regress after 14 days. After a long latency (7 to 9 months) some of the mice develop progressively growing tumors. Also, rare anchorage-independent variants of the cells arise after further growth of the cultures, and when colonies from agar are isolated and cells injected into nude mice, large (>2.5 cm) tumors form which do not regress. Thus, the *ras* oncogene initiated a multistep carcinogenic process in these cells. The first detectable change was an alteration in differentiation which was followed by escape from senescence, acquisition of anchorage-independent growth ability, and ultimately tumorigenicity.

These results seem to differ from the findings discussed above with mouse keratinocytes and rat tracheal cells where it was concluded that chemically induced alterations in growth and/or differentiation of these cells are most likely not due to mutation of the *ras* gene, based on the frequency of chemically induced changes and the inability of viruses with the *ras* oncogene to induce expanding colonies of differentiation-altered mouse keratinocytes. The basis for this species- or cell-specific difference is still unclear, and more experiments of this type with epithelial cells are needed. Since human cells including human epithelial cells are difficult to transform with chemical carcinogens (as discussed in greater detail later in this chapter), it is unlikely that mutation of the *ras* oncogene is sufficient to initiate the transformation of these cells in culture. The results of Yoakum et al.[105] do indicate that the *ras* oncogene has an important effect on human cells. The long time period and multiple selection methods required for *ras*-transfected bronchial cells to express tumorigenic potential indicate that multiple events are necessary. Therefore, mutation of the *ras* gene may be one key event, but this gene may not be the only target for carcinogens. Therefore, it is possible that even though *ras* activation may occur as the result of carcinogen treatment, other carcinogen-induced changes also occur, and these are more likely to result in the changes most readily observable in vitro (i.e., enhanced growth); hence, these cells are selected in culture, and *ras* gene activation then becomes a later event in the neoplastic progression of these cells.

This raises a critical point which must be remembered in understanding in vitro models of carcinogenesis. Often by necessity, one of the earliest changes which must occur in order to study the process of neoplastic development in vitro is escape from senescence. If the cells senesce, progression to malignancy cannot be studied in culture. Therefore, acquisition of immortality has to be an early event in transformation of cells in culture, but this may be only a tissue culture artifact and have no relevance to the growth of cells in vivo. On the other hand, immortality may be a critical event in the neoplastic process of cells in vivo as well as in vitro. In the next section, therefore, we will describe in more detail what is known about this phenotypic change and discuss the evidence for and against its role in carcinogenesis.

C. Immortality

When cells are cultured from normal embryonic or adult tissue under appropriate conditions, rapid cell division occurs, and the cells can be subcultured for a varying number of passages. The rate of cell division remains constant for several weeks or months, but after a certain number of cell divisions or population doublings, the rate of cell growth decreases, and ultimately the vast majority of the cells cease proliferating. This phase of cell growth is termed crisis, and the cells in crisis undergo various morphological and degenerative changes. The limited capacity to divide in the life span of normal cells has been referred to as cellular aging or senescence.[107] Certain cell populations acquire an indefinite life span or division potential in culture, and this escape from senescence is termed "immortality". Certain tumor-derived cells are immortal.[107] Cells from normal tissue will escape senescence and become immortal following treatment with carcinogenic chemicals or viruses. Immortalization of some cell cultures also occurs spontaneously.

Two major factors, culture conditions and species of origin, affect the life span of cells in culture and the frequency of spontaneous immortalization. Culture factors, i.e., media, serum, and serum albumin, affect the number of population doublings of fibroblasts in culture.[43,107-109] However, even under optimal culture conditions, normal cells have a finite life span. The number of population doublings prior to crisis varies among species. Human cells generally grow for 50 to 70 population doublings before senescence. Extensive attempts to increase this life span by varying culture conditions have failed.[107] The aging of fibroblastic cells in culture has been studied in greatest detail in part due to the fact that the early tissue culture methods favored growth of fibroblasts. However, with the development of optimal growth conditions for human epithelial cells, it has been shown that these cells can also be grown for a limited number of population doublings.[110] Rodent cells generally grow for fewer population doublings (10 to 30) before crisis.[43,108,109]

The frequency of spontaneous immortalization varies considerably between cells of different species. Human cells very rarely, if ever, escape senescence spontaneously.[43,107] In contrast, immortal cell lines can be obtained from cultures of mouse cells 100% of the time.[43,111] Hamster and rat cells are intermediate, and the frequency of establishment of immortal cell lines with these cells varies with the culture conditions used.[12,43,112] This is likely due to the increased probability as a function of time or growth in culture for the random occurrence of a spontaneous heritable alteration resulting in immortalization.[113] The intrinsic rate per generation of this putative change must vary considerably between cells of different species to account for the striking differences in terms of spontaneous immortalization between human and rodent cells.

Although induction of immortality by chemicals cannot be shown with mouse fibroblasts due to the high spontaneous frequency of this change in these cells,[43,44,111] carcinogens clearly enhance the immortalization process of other rodent cells.[12,25,27,36-38,83] This effect of carcinogens was recognized for many years, but it was not emphasized by some authors because it was not clear whether it was a key event in the pathway of neoplastic development or just an in vitro epiphenomenon. The studies of Newbold and co-workers[27] were the first to provide concrete evidence that immortality may be a prerequisite for neoplastic transformation. These authors demonstrated that carcinogen treatment of Syrian hamster cells induces rare, immortal variants which are not initially anchorage independent, but the progeny of which frequently progress to anchorage independence after further growth. These investigators were able to isolate anchorage-independent Syrian hamster cells which are not immortal; these cells are also nontumorigenic. Conversely, immortal, nonanchorage-independent cells were isolated and also are not tumorigenic. However, anchorage-independent variants of immortal cells are tumorigenic. These results support the hypothesis that immortality, although insufficient by itself, may be a prerequisite for neoplastic transformation.[27] This conclusion was further supported by transfection experiments with oncogenes. By this technique, it was shown that the mutated *ras* oncogene can induce anchorage independence[19] and tumorigenicity[21] of immortal Syrian hamster cells, whereas similarly treated normal, diploid Syrian hamster cells may express morphological changes and anchorage-independent growth while retaining a finite life span and lack of tumorigenicity.[19,21]

Further support for the importance of immortality was provided by experiments which demonstrate that normal cells become immortal following infection or transfection with viruses or DNA containing certain oncogenes, including *myc,* N-*myc, myb,* adenovirus Ela, polyoma large T, SV-40 large T, Epstein Barr virus DNA, and *p53*.[18,20,56,57,59,114-118] These oncogenes share the common property of being associated with the nucleus.[59] Several of these oncogenes (*myc,* N-*myc,* Ela, polyoma large T, SV-40 large T, and *p53*) also cooperate with the *ras* oncogene in the neoplastic conversion of primary cells.[18,59,119,120]

The biochemical basis for the action of these oncogenes in immortalization is a key, but still not completely understood, event. Transfection of rat fibroblast cells with *myc* or polyoma large T reduces their serum requirement for growth,[56] indicating either the abrogation of a growth factor requirement or a hypersensitivity to low concentrations of growth factors. The *myc* gene is an inducible gene that is regulated by specific growth signals such as platelet-derived growth factor (PDGF) in a cell cycle-dependent manner,[121] and the c-*myc* gene under the control of the mouse mammary tumor virus promoter or the c-*myc* protein upon microinjection can partially relieve the requirement of fibroblasts for PDGF.[122,123] The v-*myc* oncogene can also abrogate the requirement of hematopoietic cells for interleukin-3 or -2 growth factors[124] and imparts a hypersensitivity to epidermal growth factor[125] to chicken heart mesenchymal cells.

These results strongly implicate *myc* in regulation of growth and provide a logical basis for its immortalizing ability. The exact mechanism(s) involved may depend on the specific cell type involved. Epstein-Barr virus-immortalized B lymphocytes release a soluble factor which mimics B-cell mitogenic factors.[126] Therefore, other genes act by stimulating ectopic growth factor production.

While it is clear that the nuclear oncogenes play a role in growth control processes,[59] direct evidence linking these effects to immortalization is often lacking. It is possible that immortalization is not directly caused by the action of these oncogenes, but is the consequence of the enhanced proliferation of the treated cells. *Myc*-containing viruses cause a dramatic stimulation of cell proliferation.[127] The immortalization ability of *myc* has been demonstrated with rat fibroblasts which have an intrinsic, spontaneous rate of immortalization.[43,113] Enhanced proliferation of these cells by oncogenes may simply increase the probability of spontaneous immortalization. The major difficulty in interpreting these experiments is that it is impossible to identify an "immortal cell"; immortality is a population property, not a cellular attribute. The only quantitative assay used for immortalization is the ability of the cells to grow at low cell density.[18,56] This assay clearly does not measure immortalization but rather enhanced proliferative ability. Cells demonstrating this growth enhancement become immortal,[56] but this may be a secondary event. Oncogene-induced immortal cells are sometimes aneuploid,[115] and aneuploidy induction has been proposed as a possible mechanism for carcinogen-induced immortality.[128,129] With fibroblasts, no attempts have been made to determine the proportion of cells in a *myc*-transformed colony which goes on to become immortal. In the case of *p53*-treated cells, only a small fraction of transformed cells is capable of continuous growth.[120] Furthermore, *p53*[120] or *myc*[130] alone can make certain immortal cell lines tumorigenic. These results indicate that these oncogenes can participate in the neoplastic transformation process at stages other than immortalization.

In the case of the polyoma large T gene, evidence does exist for a direct role of this gene in immortalization. Rat embryo cell cultures transformed with a plasmid containing a temperature-sensitive mutant of the large T yield established cell lines at the permissive temperature; these cells are growth arrested when shifted to the nonpermissive temperature, and a progressive loss in cell viability ensues.[20] These results indicate a continuous requirement for large T function for maintenance of immortality. Unfortunately, no temperature-sensitive mutants exist for other nuclear oncogenes.

Carcinogen-induced immortalization has been studied in a variety of cellular systems. Newbold et al.[27] demonstrated that several carcinogens induce immortalization of Syrian hamster cells. The frequency of carcinogen-induced immortal cells is low (10^{-7} to 10^{-5} per viable treated cell). These results are consistent with a mutational change in a single gene. Since certain oncogenes, i.e., *myc* and *p53*, are known to be activated by rearrangements,[59,131] these results are consistent with a mutational activation of an immortalizing gene.

The frequency of immortal cell variants is considerably lower than the frequency of carcinogen-induced morphologically transformed Syrian hamster embryo cells. As discussed earlier, morphological transformants occur at frequencies of 10^{-2} to 10^{-3}, at least three orders of magnitude larger than immortal cell variants reported by Newbold et al.[27] Since morphologically transformed colonies become immortal, these changes have been considered related. Several explanations can be offered for the differences in frequency of these two phenotypic transformants.

1. Not all morphologically transformed colonies can be established into cell lines; thus, targets other than immortalization genes may be activated in morphologically transformed cells, and only a subset of the morphologically transformed colonies may have mutated immortalization genes. This explanation certainly has some applicability; however, since ≥10% of the morphologically transformed

mechanism and suggested that an epigenetic process is involved in the carcinogen-induced transformation of C3H 10T1/2 cells. Based on further analysis, this group[184] later proposed that the second step behaves like a spontaneous mutation with a constant, but small, probability of occurring at each cell division.

Fernandez and co-workers[183] also have made similar observations with the C3H 10T1/2 cells and have proposed a "probabilistic theory" to explain the formation of transformed foci by these cells following 3-methylcholanthrene treatment. Their theory is similar to that proposed by Kennedy et al.[182] in that two steps must occur for cell transformation. The first step is the "activation" of a large percentage of the cells by the carcinogen, which occurs with a probability p_1, and the second step is the transformation of the activated cells, which occurs with a probability p_2 per cell generation. The authors have derived a mathematical equation which predicts the frequency of focus formation based on the probability of these two steps (p_1 and p_2) plus the probability of deactivation per cell generation of the carcinogen-activated cells, which is termed p_3.

Alternative explanations for the results with C3H 10T1/2 cells have been suggested. Mordan et al.[185] have suggested that at least part of the difficulty in quantitating focus formation of C3H 10T1/2 cells is due to the suppressive effects of normal cells on the expression of foci formation by the transformed cells. These authors suggest that a minimum colony size of ~128 transformed cells at confluence is required for the formation of a transformed focus. The suppression or reversion of morphological transformation of C3H 10T1/2 and BALB/c 3T3 cells has been reported previously.[186,187] Haber et al.[181] originally suggested that carcinogens affect two parameters in C3H 10T1/2 transformation. The first was the induction of the potential for transformation which occurred in a large percentage of these cells. This is analogous to the activation step proposed by Fernandez et al.[183] This induction occurs in nearly all of the cells, and according to Haber et al. is not dose dependent. The second effect of carcinogen treatment suggested by Haber et al. is to influence cell-cell interactions in a dose-dependent manner to allow for the expression of the transformed potential of the cells.

At present, it is apparent that the expression of morphological transformation of C3H 10T1/2 cells is not a one-step process. The first step appears to be rapid event[188] that occurs in a high percentage of the cells.[181-183,189] The second step could either be a qualitative change in the cells that occurs at a low frequency during the growth of the cells or at confluence, or this second step could be amplification of the transformed cells to overcome the suppressive effects of the nontransformed cells. Both mechanisms may be operative. Further experiments are needed to elucidate the mechanism of transformation with C3H 10T1/2 and the relevance of carcinogen-induced events in these cells to neoplastic progression in vivo.

As discussed earlier, rat embryo cells undergo spontaneous establishment into cell lines relatively easily; however, the immortal cell lines which arise require considerable time in culture before spontaneous tumorigenic conversion occurs. Treatment of immortal rat embryo cell lines with carcinogens induces neoplastic transformation of these cells but only rarely.[190] Fisher rat embryo cell lines with fewer than 60 population doublings are resistant to both spontaneous and chemically induced neoplastic transformation, whereas the same cells chemically infected with the nontransforming retrovirus Rauscher leukemia virus are sensitive to transformation by a variety of chemical carcinogens.[191-194] These results suggest that multiple events are necessary for the transformation of these cells.

Thomassen and DeMars[195] have quantified the spontaneous development of anchorage independence and tumorigenicity in a near-diploid line of morphologically transformed mouse cells (CAK). Anchorage-independent, nontumorigenic variants of CAK cells originate spontaneously at an estimated rate of about 10^{-4} per cell per generation. Tumorigenic variants appear spontaneously during proliferation of an anchorage-independent cell clone at an estimated rate of about 10^{-7} per cell per generation, but are undetectable among anchorage-dependent CAK cells. In contrast, MNNG treatment induces the appearance of tumorigenic variants in both anchorage-dependent and independent clones with an estimated frequency of about 10^{-4} per surviving clone in the former, which is similar to the induced frequency of ouabain-resistant variants in the same cells. Anchorage independence is expressed without tumorigenicity in new anchorage-independent variants, but tumorigenic cells are always anchorage independent. These authors proposed that CAK cells become tumorigenic by a three-step pathway that includes changes causing morphological transformation, anchorage independence, and tumorigenicity. The evidence is also consistent with an alternative two-step pathway where anchorage independence and tumorigenicity are acquired in a single step, since anchorage-independent, tumorigenic clones are derived from anchorage-dependent cells soon after a single mutagenic treatment.

A number of hamster cell lines have been established and studied in many laboratories.[39,196,197] Diamond[196] reported that two sublines, Nil-1 and -2, from the same primary culture become tumorigenic after widely different numbers of passage. Another established cell line derived from the kidney of a newborn Syrian hamster, designated BHK for baby hamster kidney, has been widely studied for many years.[197] Bouck and DiMayorca[198] have presented evidence for somatic mutation as the basis for "malignant" transformation of BHK cells based on the following observations: BHK cells have a low frequency of spontaneous transformation for the ability to grow in soft agar; the frequency of this phenotype is increased by treatment with mutagens; this phenotype is stable and has a low frequency of reversion; and the phenotype is often temperature sensitive, which suggests that it arises from a temperature-sensitive gene product, typically a protein derived from a missense mutation. Although a mutated oncogene has not been identified in chemically transformed BHK cells, these results are consistent with this hypothesis. The single-step mutational hypothesis for BHK transformation has been criticized,[60,199] but this is still the best explanation for the observations with this cell line.

The results with another Syrian hamster cell line,[39] FOL-2, are quite different from the studies with BHK cells. The frequency of spontaneous growth in agar of FOL-2 is less than BHK cells. Crawford et al.[200] and Crawford[202] measured the spontaneous transformation rate to growth in agar by a Luria-Delbrück fluctuation test. They observed that transformants that grow in agar arise at a rate (0.57 to 5.5×10^{-7} transformants per cell per generation) which is consistent with a specific locus gene mutation in these cells (7.8×10^{-8} Oua^R mutants per cell per generation). However, several experiments have demonstrated that the frequency of transformation of FOL-2 cells is not increased by treatment with mutagens.[201,202] Thus, despite the apparent similarity of the spontaneous rates of transformation and gene mutation in these cells, these two processes are clearly distinct in terms of their sensitivity to mutagen treatment. Crawford et al.[200] have proposed that the transformation and neoplastic conversion of FOL-2 cells may be the result of chromosome segregation. This hypothesis is supported by their studies on the rate of nondisjunction in aneuploid Syrian hamster cells and the behavior of cell-cell hybrids which suppress or express anchorage-independent growth and tumorigenicity depending on chromosome segregation.[202]

The neoplastic conversion of immortal Chinese hamster cell lines has been studied in several laboratories. Sager[203] and Sager and Kovac[204] have extensively examined the mechanism(s) of neoplastic progression of Chinese hamster embryo fibroblast (CHEF) cell lines. Two sublcones, CHEF 16-2 and CHEF 18-1, established from a male Chinese hamster embryo by Yerganian, have been studied.[204] Both clones are permanent cell lines with high cloning efficiencies on plastic and have a diploid (apparently euploid) karyotype. However, they differ greatly in terms of their transformed properties. CHEF 16-2 cells are highly tumorigenic in nude mice, producing tumors with as few as 10 cells in 8 to 10 weeks, while CHEF 18-1 cells are nontumorigenic after injection of as many as 10^7 cells. CHEF 16-2 cells also have an altered colony morphology and grow in methylcellulose and agar or on plastic in medium with low serum (1%) concentration. CHEF 18-1 cells do not have any of these transformed cell properties and remain stable and nontransformed after many passages in culture. CHEF 18-1 cells also fail to produce tumors if implanted in vivo attached to a substrate in contrast to other immortal cell lines. These findings indicate the CHEF 18-1 cells are very similar to early-passage normal, diploid cells although they have transformed into an immortal cell line with an increased cloning efficiency.

Even though CHEF 16-2 cells are highly tumorigenic, Kritchin and Sager[205] have proposed that these cells are premalignant. This conclusion is based on the observation that the cells undergo extensive karyotypic changes in vivo during clonal establishment as a tumor. Loss of chromosome stability, leading to rapid genomic rearrangement and possibly mediated by transposition, has been suggested as the mechanism of this late-stage transformation of CHEF 16-2 cells in vivo.[203,205]

These authors have also examined the induction of neoplastic progression of CHEF 18-1 cells.[206] Treatment of the cells with mutagenic chemicals (e.g., MNNG, EMS [ethyl methane sulfonate], and 4-NQO) at doses which caused 50 to 90% cell killing, resulted in a low frequency of tumorigenic cells, but a high frequency of the surviving cells that can grow in methylcellulose (1 to 4.6×10^{-3}) or in low serum (0.4 to 1.4×10^{-2}). These variant cells occur at a higher frequency than gene mutations measured in the same cells (induced mutant frequencies = 1 to 7×10^{-5} per surviving cell).[206] Some, but not all, of the anchorage-independent and low serum "mutants" have a higher probability than the parent CHEF 18-1 cells to subsequently undergo neoplastic transformation. The anchorage-independent mutants do not grow in low serum; however, most of the low serum mutants are anchorage independent.

Three of seven anchorage-independent clones are tumorigenic in nude mice, and the number of tumors increases after a second mutagenic treatment of these clones. The other four anchorage-independent clones are nontumorigenic in nude mice even after mutagenesis. One of the four low serum clones is tumorigenic in nude mice, and the other three clones are tumorigenic after secondary mutagenesis. Only one of eight randomly isolated clones surviving an initial mutagenic treatment, but not selected on the basis of a transformed phenotype, is tumorigenic after secondary mutagenesis compared to six of ten clones selected on the basis of anchorage-independent or low serum growth. This suggests that mutagenesis per se does not increase the frequency of tumorigenic clones. Cells isolated as having both anchorage-independent and low serum growth phenotypes are not tumorigenic (none of seven clones) indicating that these two phenotypes are not sufficient for tumorigenicity.

In conclusion, the authors propose that neoplastic transformation of CHEF 18-1 cells is a multistep process. Anchorage-independent or low serum mutants of the parent cells can be induced at high frequencies by mutagens. The acquisition of these characteristics is not sufficient for tumorigenicity, but increase the probability of further progression. This probability is increased by additional mutagenic treatments. Heterogeneity in the propensity for progression exists for at least the anchorage-independent mutants. At least two complementation groups for this phenotype have been identified in these cells.[207] The intermediate, preneoplastic stages of anchorage-independent and low serum growth may be necessary, but not sufficient, changes for tumorigenicity or may enable the CHEF 18-1 cells to survive in vivo and thereby facilitate the appearance of new alterations needed for tumorigenicity. These hypotheses may be interrelated in that cells with mutant phenotypes may have a selective growth advantage, and further genomic rearrangements may result in the evolution of new phenotypes required for malignancy.[206] Thus, Sager[203] has described the carcinogenesis process as "speeded-up or accelerated evolution".

Kraemer et al.[208] have also studied spontaneous neoplastic evolution of Chinese hamster cells in culture. These authors were able to divide the process into multiple steps. After the cells escape crisis, multiple steps are required for complete expression of the tumorigenic state. Thomassen et al.[209] quantitated the rate of spontaneous generation and the frequency of carcinogen-induced anchorage-independent variants of immortal RTE cells in culture. Anchorage-independent variants of different RTE cell lines arise spontaneously by a stochastic process at rates of 0.5×10^{-4} to 5.4×10^{-4} variants per cell per generation, as determined by fluctuation analyses. These variants are also induced by the mutagen N-methyl-N'-nitro-N-nitrosoguanidine with a frequency of $\sim 10^{-3}$ variants per surviving cell. The rates of spontaneous change and the frequencies of induction are similar to some, although not all, rates and frequencies of change to anchorage independence for fibroblast-like cells in culture. In addition, these rates and frequencies are similar to those for mutations at some known gene loci. The induced frequency of this late change in neoplastic progression is, however, considerably lower than the frequency of induction of the initial, preoplastic changes in RTE cells in culture ($\sim 3 \times 10^{-2}$ per surviving cell) suggesting that different mechanisms may be involved in early vs. late stages of neoplastic progression of these cells.

When a comparison is made of studies measuring the spontaneous rates and induced frequencies of anchorage-independent growth of different immortal cell lines in culture, the most striking observation is the heterogeneity of response.[209] For example, the calculated rates of acquisition of anchorage-independent growth vary from 10^{-7} per cell per generation to 10^{-4} per cell per generation.[209] This heterogeneity is inconsistent with the hypothesis that the conversion of immortal cells to the tumorigenic state requires only the activation of a single oncogene. A more likely explanation, which is supported by other evidence discussed below, is that multiple changes are required at least for some cell lines and that oncogene activation is necessary but not sufficient for neoplastic transformation of immortal cells.

B. Possible Mechanisms for Late Stages of Neoplastic Development of Cells in Culture
1. Oncogenes

It is well established now that oncogenes are involved in the tumorigenic conversion of immortal cells in culture. Transforming genes have been identified in a number of cells transformed in vitro by chemicals or radiation. Shin and colleagues showed that 3-methylcholanthrene-transformed 10T1/2 cells contain a gene capable of transforming NIH/3T3 cells following DNA transfection. This was later shown to be an altered Kirsten ras gene.[210] Radiation-induced transformed 10T1/2 cells also have transforming genes, but the specific oncogenes involved have not been identified.[24]

Fetal guinea pig fibroblasts transformed by treatment with four different chemical carcinogens including nitroso compounds and polycyclic hydrocarbons also have activated oncogenes which transform NIH/3T3 cells following DNA transfection.[42,43] In this case, an activated Ha-ras gene is involved. The same gene is activated in each cell line independent of the carcinogen used initially. Furthermore, the mutation of the oncogene was a late event in the transformation process since DNA from preoplastic cell lines did not transform NIH/3T3 cells.[212,213]

Other in vitro transformed cells have been shown to contain activated transforming genes. Syrian hamster embryo cells appear to have both *ras*- and non-*ras*-related oncogenes activated.[211,214,215] Chinese hamster (CHEF-16) cells have activated transforming genes which are detected following transfection into a nontumorigenic Chinese hamster cell line (CHEF-18).[216]

Further evidence for a role of oncogenes in late-stage events in neoplastic transformation comes from studies of transfection of known oncogenes into nontumorigenic, immortal cell lines. Immortal cell lines from the mouse,[217,218] rat,[17,18,59,219] Chinese hamster,[220,221] and Syrian hamster[19,21,40,221] become tumorigenic after transfection with different transforming oncogenes, i.e., *ras* or *src*. However, in some cases immortal cell lines are either refractory to transformation by oncogenes or neoplastic transformation of immortal cells requires multiple steps or multiple oncogenes. For example, Ruley et al.[219] have shown that unlike Rat-1 cells, another established rat cell line, REF-52, is not transformed by the Ha-*ras* oncogene alone. The adenovirus early gene E1a together with the Ha-*ras* gene converted REF-52 cells to the neoplastic phenotype. These findings indicate that cellular immortality is not a sufficient prerequisite for neoplastic transformation by the *ras* oncogene. Additional steps beyond immortality and activation of a transforming gene seem to be required. Tsunokawa et al.[222] observed that whereas colonies of NIH/3T3 cells expressing v-Ha-*ras* showed transformed phenotype colonies, another immortal mouse cell lines (m5S) with a functionally active, integrated v-Ha-*ras* gene did not show these changes, suggesting that *ras* expression alone does not necessarily transform immortal murine cells.

Jenkins et al.[131] have shown that cells immortalized by a mutant of *p53* gene were refractory to subsequent transformation by the *ras* oncogene. Interestingly the wild-type *p53* gene induced immortal cell lines which were sensitive to transformation by *ras*. These results suggest that immortalization and *ras* cooperation are separate activities indicating that changes in addition to immortalization and transformation are involved.

Thomassen et al.[21] studied the neoplastic transformation of an immortal Syrian hamster cell line following transfection with v-Ha-*ras*. They observed that all clones which expressed v-Ha-*ras* DNA were tumorigenic; however, the clones were highly variable in terms of their latency periods in vivo and anchorage-independent growth. Neither of these two parameters correlated with the level of expression of v-Ha-*ras* RNA. All the cell lines derived from tumors and reinoculated into nude mice had short latency periods in vivo, were highly anchorage independent, and had high levels of v-Ha-*ras* expression. These results suggest that, in these experiments, v-Ha-*ras* expression was necessary, but not sufficient, for the tumorigenicity of these cells and that additional changes in the cells were acquired.

These results support the conclusion that oncogenes are involved in the late stages of neoplastic transformation of cells in culture, but also indicate that additional changes are required. The observations that two cooperating oncogenes are necessary to neoplastically transform normal, primary cells support the hypothesis that only two steps (immortalization and transformation) are necessary for tumorigenic conversion.[17,18] In order to determine whether these two steps are sufficient for neoplastic development, Oshimura et al.[22] performed cytogenetic analyses of tumors which formed after transfection of Syrian hamster embryo cells with *ras* plus *myc* oncogenes and found that they had a nonrandom chromosome change, monosomy of chromosome 15. The tumors were also monoclonal in origin. These results suggest that *ras* plus *myc* is necessary but not sufficient for neoplastic transformation of these normal cells. An additional change, loss of chromosome 15, is required or advantageous for tumorigenicity induced by v-Ha-*ras* plus v-*myc* oncogenes. Thus, the neoplastic progression of normal, diploid cells requires more than two steps under certain conditions.[22]

2. Tumor Suppression Genes

The identification of activated oncogenes in tumors by DNA transfection techniques requires that these genes are dominant, allowing expression of the neoplastic phenotype to occur. In contrast, cell hybrids between tumor and normal cells are often nontumorigenic indicating that tumorigenicity is recessive.[220,223] This paradox can be explained in several different ways: (1) during DNA transfection and selection for the transformed cell, some normal genes may be lost allowing expression of the neoplastic phenotype; (2) the ability of a cell to suppress the tumorigenic phenotype may be dependent on the stage of progression of the cell; (3) the ability to suppress the tumorigenic phenotype may depend on the oncogenes activated; and (4) the expression of the neoplastic phenotype may depend on the dosage of the putative suppressor and transforming genes.

These hypotheses, which are not mutually exclusive, are each supported by certain studies. The nonrandom loss of chromosome 15 in tumors of Syrian hamster embryo cells induced by *ras* plus *myc* transfection[22] may be an example of the first mechanism. This conclusion is supported by studies of hybrids between *ras/myc* tumor cells and normal Syrian hamster embryo cells.[224] The tumorigenicity of cells derived from *ras/myc* tumors is reduced in hybrids with normal cells. When these hybrid cells are passaged, anchorage-independent variants appear in the cultures which are accompanied by morphological changes and tumorigenicity. These transformed segregants of the hybrid cells have a nonrandom loss of chromosome 15 supporting the conclusion that the loss of chromosome 15 results in the loss of a cellular gene which affects a phenotype change necessary for neoplastic development.

The second mechanism listed above is indicated in studies of the loss of tumor suppression function during chemical carcinogen-induced neoplastic progression of Syrian hamster embryo cells. Koi and Barrett[225] have shown that tumorigenicity and anchorage-independent growth are suppressed in cell hybrids between normal, early-passage Syrian hamster embryo cells and a highly tumorigenic, chemically transformed hamster cell line, BP6T. These two phenotypes segregated coordinately in these cells. To determine at what stage in the neoplastic process this tumor suppression function was lost, two chemically induced, immortal cell lines were examined at different passages prior to neoplastic conversion for the ability to suppress the tumorigenic phenotype of BP6T cells following hybridization. Hybrids of BP6T cells with the immortal, nontumorigenic cell lines at early passages did not express tumorigenicity or anchorage-independent growth. At later passages prior to neoplastic transformation, this tumor suppression ability is reduced in the same cells and in some cases nearly completely lost. Subclones of the cell lines are heterogeneous in their ability to suppress tumorigenicity in cell hybrids; some clones retain the tumor suppression ability and others lose this function. The susceptibility to neoplastic transformation of these cells following DNA transfection with the viral *ras* oncogene or BP6T DNA inversely correlates with the tumor suppressive ability of the cells. Based on these results the authors suggest that chemically induced neoplastic progression of Syrian hamster embryo cells involves at least three steps: (1) induction of immortality, (2) activation of a transforming oncogene, and (3) loss of a tumor suppression function.

The ability to suppress the tumorigenic phenotype may depend on the oncogenes activated in the tumor cells (mechanism 3). For example, *ras*-transformed cells are suppressed in hybrids with normal cells even though the *ras* gene is expressed in the hybrids;[220] in contrast, cells transformed by DNA viruses are sometimes not suppressed by hybridization with normal cells.[223,226] Evidence for a dosage effect of oncogenes and suppressor genes (mechanism 4) is indicated in certain studies. For examples, hybrids between a near-diploid, human tumor cell (HT 1080) and normal diploid fibroblasts are suppressed for tumorigenicity;[227] however, when near-tetraploid variants of the same tumor cell are fused with normal fibroblasts, the resultant hybrids are tumorigenic suggesting that tumorigenicity in these hybrids is affected by gene dosage.[227]

Studies on recessive cancer genes in human tumors (for example, retinoblastoma and Wilms' tumor) also support the hypothesis that a normal cellular gene, termed a tumor suppressor gene, has to be lost or inactivated for the development of these tumors.[228,229] Loss or inactivation of this gene appears to be an additional step in the process of neoplastic development in cells in culture and tumors in vivo.

3. Differentiation Defects and Neoplastic Transformation of Cells in Culture

An understanding of the biological role of tumor suppression genes in carcinogenesis and normal cellular functions of these genes must await cloning and identification of these genes. One possible role for these genes is in the normal processes of differentiation. Recent studies by Scott and colleagues,[230] Scott and Maercklein,[231] and Willie et al.[232] have indicated that defects in the control of cell proliferation and differentiation play an important role in neoplastic transformation of cells in culture. These investigators have studied an immortal cell line, 3T3-T, which retains the ability to differentiate in culture into adipocytes. Differentiation occurs when the cells are in a distinct G_1 arrest state, G_D. Cells in G_D can be induced by various factors to express differentiative features including terminal differentiation. These cells also are arrested in other distinct G_1 states by growth factor deficiency or nutrient deficiency. Differentiation is not manifested in cells arrested in these G_1 states.[230] To test the hypothesis that carcinogenesis is associated with a defect in the control of cell differentiation, they tested the effects of low doses of a carcinogen, UV irradiation, on the ability of 3T3-T cells to differentiate. It was observed that UV irradiation selectively and stably inhibits the differentiation of a high percentage of proadipocyte stem cells without altering their ability to regulate cellular proliferation in other G_1 states, i.e., in growth factor- or nutrient-deficient media. The UV-induced differentiation-defective cells are not completely transformed, but have an increased rate of spontaneous transformation.[231] These results show that transformation of immortal 3T3-T cells is a multistep process, and one step involves a defect in the control of differentiation. Whether or not this is related to the loss of tumor suppression function is an interesting question which deserves further research.

4. Promotion of Cell Transformation

The enhancement of transformation by known tumor promoters has been reported with a variety of cell transformation assays.[13,14,233,234] Both normal, diploid Syrian hamster embryo cells and immortal cell lines including C3H 10T1/2 cells, BALB/c 3T3 cells, a mouse epidermal cell line, and virally infected cells have shown an enhanced response following treatment with phorbol ester promoters. In the case of the 10T1/2, 3T3, and Syrian hamster embryo cells, tumor promoters, i.e., 12-O-tetradecanoyl-phorbol-13-acetate (TPA), are virtually inactive or weakly transforming alone, but significantly enhance the frequency of transformation if the cultures are pretreated with a low dose of carcinogen.[13,14,234] In contrast, TPA alone induces significant transformation of JB-6 mouse epidermal cells or adenovirus-infected rat embryo cells.[235,236] These cells are considered to be initiated prior to promoter treatment either by the carcinogen used to induce immortality in the case of JB-6 cells or by the adenovirus in the case of the rat embryo cells. With Syrian hamster embryo cells, 10T1/2, and BALB/c cells the enhancement of carcinogen-induced morphological changes is usually initially reversible.[14,233,234] In contrast, TPA alone induces an irreversible change in anchorage-independent growth of JB-6 cells and adenovirus-transformed rat embryo cell lines.[235-239] No definitive proof for the mechanism of action of promoters exists for these systems, although several possible mechanisms have been proposed.[14,234] The different responses of each system would indicate the possibility of multiple mechanisms.

TPA also enhances the transformation of normal and immortal cell lines transfected with oncogenes.[240-242] The enhancement by TPA of immortal cell lines transfected with ras gene[240] supports the hypothesis that multiple steps are required for late stages in cell transformation.

Further studies are required to understand enhancement of cell transformation by promoters. These studies may yield important insights into the mechanisms of carcinogenesis. However, it should be cautioned that the changes induced by promoters in vitro may be unrelated to the essential changes necessary for tumor promotion in vivo. Most cell culture models of tumor promotion were developed using TPA as a model compound. The effects of TPA in cells in culture are multiple, and it is possible that the observed "promotion" by TPA has no relevance to the in vivo process of tumor promotion.

5. Single-Step Neoplastic Transformation and Other Inconsistencies with the Immortalization/Transformation Model

In the preceding discussion we have stressed the multistep nature of neoplastic transformation. There are, however, certain oncogenes or oncogenic viruses which appear to induce tumors by causing a single-step neoplastic transformation. These cases deserve additional discussion.

Spandidos and Wilkie[221] reported that when the EJ-ras gene is linked to two transcriptional enhancers, neoplastic transformation of early passage rat or Chinese hamster cells can be induced by ras alone without a second oncogene. These results may indicate that a single-step transformation by ras alone is possible. However, fully transformed malignant cells obtained after transfection of early-passage Chinese hamster cells with the mutated ras with transcriptional enhancers possess radically altered karyotypes.[243] Therefore, the possibility that additional changes occurred in these cells seems probable, and the number of steps involved in the transformation of these cells cannot be determined.

Yoakum et al.[105] have reported that transfection of Ha-ras oncogene alone into human bronchial epithelial cells can result ultimately in neoplastic transformation of the cells. However, this is clearly a multiple step process since the cells are not initially tumorigenic after transfection, and many months and multiple selective procedures for cells at different stages of the process are required to isolate tumorigenic cells. Spandidos and Wilkie[221] also reported that ras alone can immortalize early passages of Chinese hamster cells. Therefore, the hypothesis that ras only transforms immortal cells is complicated by these findings. A logical explanation for these results is that ras acts as a stimulator of cell proliferation. In some cases this can result in immortalization, probably as a consequence of this stimulus.

The two-step model of tumorigenic conversion, involving immortalization and transformation where each step is controlled by a specific class of oncogenes, is also complicated by the observations that the "immortalizing genes", myc and p53, are able to neoplastically convert immortal cell lines in some experiments.[120,130] These results should not be interpreted as disproving the general paradigm. As we discussed earlier the process of neoplastic conversion is probably more complicated than the two steps of immortalization and transformation. Results of many laboratories with oncogenes and chemicals indicated that these two steps are important. Given the complexity of the process and the fact that the myc and p53 are known to alter growth responses of cells, the observations that these genes can influence the tumorigenic process in different stages does not seem surprising.

The results with DNA transfection of cells in culture should be compared with the induction of tumors in vivo by oncogenic viruses carrying the same oncogenes. For example, the inability or a weak activity of the Ha-ras oncogene to transform cells in culture following transfection is distinctly different from the results in vivo and in vitro with Harvey murine sarcoma virus. This virus rapidly induces tumors in vivo within a few weeks and causes morphological and neoplastic transformation of mouse, rat, and hamster embryo cells in vitro.[244] However, the v-Ha-ras and EJ-ras oncogenes only cause a transient morphological change in primary embryo cells following DNA transfection.[17,19,21] The role of virus infection, replication, and helper virus in transformation must be examined before direct comparisons can be made with DNA transfection results. Bather et al.[245] reported that mouse embryo cells infected in vitro with a murine sarcoma virus are not immortal, since they fail to undergo sustained cell division necessary to form foci; focus formation appears to result from virus release and reinfection of neighboring cells. In contrast, transformation of cells by ras oncogene by transfection occurs in the absence of virus replication and reinfection of untransformed cells. Since the oncogene alone can induce a transient transformation to anchorage independence and altered morphology,[17,19,24] this transient transformation may be sustained by recruitment of newly transformed cells. It is possible that the continued proliferation of partially transformed cells allows additional changes necessary for immortality and stable transformation to occur spontaneously.

Similarly, Rous sarcoma virus (RSV) can induce morphological and neoplastic transformation of primary chicken embryo fibroblasts.[237] The major mechanism contributing to this type of transformation appears to be viral replication and infection of surrounding cells.[246,247] Experiments by Cooper and Okenquist[248] showed that stable transformation of cultured chicken embryo fibroblasts by v-src DNA sequences only occurs in the presence of the entire RSV genome and secondary virus infection.

However, RSV can also induce tumors in mice, rats, and hamsters in the apparent absence of viral replication.[249] In addition, direct injection of a cloned, subgenomic v-src DNA fragment induces tumors in newborn chickens within 3 to 4 weeks, but all tumors regress.[247] The lack of oncogenicity of RSV DNA in vitro compared to in vivo may reflect differences in selective pressures active in vivo as compared to in vitro and further indicates that cellular changes in addition to uptake of viral oncogenic sequences are needed for tumorigenicity. Alternatively, the mode of entry of the RSV sequences in viral infection and integration as well as the effect of viral replication may play an important role in establishing the transformed state. After viral infection, integration occurs specifically in the viral long terminal repeat (LTR) sequences allowing efficient transcription of the v-src mRNA.[250] In contrast, integration of viral DNA sequences following $CaPO_4$ transfection is random and will decrease the efficiency of v-src transcription. Inefficient expression of the src gene may play a role in the failure of src to transform early passage Syrian hamster embryo cells.[40] The difference between EJ-ras alone compared to ras with transcriptional enhancers presumably also relates to a difference in expression of the gene.[221]

VI. COMPARISON OF HUMAN AND RODENT CELL TRANSFORMATION

As discussed above, it is well established that exposure of cultured normal, diploid rodent fibroblasts to chemical or physical carcinogens frequently results in transformed cells capable of forming progressively growing tumors when injected into syngeneic animals or nude mice. Although rodent cells readily undergo neoplastic transformation in response to carcinogen treatment, normal diploid human cells have proven to be extremely difficult to transform into tumorigenic cells by treatment with carcinogens.[251,252] An understanding of the differences between rodent and human cell transformation in vitro may help identify important factors and mechanisms involved in carcinogenesis. Therefore, we will discuss and compare what is known about the induction of neoplastic transformation as well as phenotypic markers of transformation in human vis-a-vis rodent cells.

A. Induction of Tumorigenicity in Human Cells

Although many phenotypic markers have been proposed to correlate with neoplastic transformation, the ability to form progressively growing tumors upon injection into an appropriate host animal is the ultimate criteria for neoplastic transformation. A vast number of reports exists showing that rodent cells transformed with chemical and physical carcinogens will form autonomously growing tumors when injected s.c. into nude mice as discussed above. Furthermore, human cell lines derived from malignant neoplasms generate tumors when injected into nude mice, many of which are serially transplantable. This indicates that the nude mouse can allow propagation of neoplastic cells of human origin. Therefore, tumorigenicity in nude mice is considered a valid marker for neoplastic transformation of both rodent and human cells.

Very few studies have reported induction of neoplastic transformation of normal human cells following treatment with carcinogens in contrast to findings with rodent cells. In 1978, Kakunaga[253] reported that treatment of cultured human fibroblasts with 4-NQO occasionally induced formation of morphologically altered foci after extended time in culture. The cell lines isolated from these foci produced progressively growing tumors when injected into nude mice. Neoplastic transformation was calculated to occur at a frequency of 1 to 3.3×10^{-7} per surviving cell. Although the extended subculturing required for expression of tumorigenicity decreases the accuracy of this estimate, this study does show that neoplastic transformation of human cells can occur at low frequencies under appropriate conditions. However, the number of reports of complete neoplastic transformation of human cells is extremely small.

The great difficulty in transforming cultured human cells is demonstrated in the extensive studies of Namba[254] where multiple treatment regimes are required to produce only a few partially transformed cells. In these studies normal human embryo fibroblasts were exposed repeatedly to Co-60 gamma irradiation with a growth period of at least two passages between treatments.[254] After 13 treatment-growth cycles (a total of 2800 rads over 40 passages), a few irradiated cultures exhibited cells with altered morphologies. These morphologically altered cells had unlimited growth capacity and abnormal karyotypes. However, none of the cells was tumorigenic in nude mice. Complete transformation of these cells to a neoplastic endpoint was not observed under conditions which would have readily transformed most rodent cells. Multiple experiments in this laboratory involving treatment of human cells with other carcinogens were also negative.

Carcinogen treatment of cultured human cells can result in cells capable of anchorage-independent growth or cells with altered morphologies.[252,255-266] However, clones isolated from these phenotypically transformed human cells are nontumorigenic or form nodules in nude mice which regress after a limited proliferation.[257]

A few investigators have studied the effects of transfecting oncogenes into cultured human cells. Sager et al.[266] reported that the EJ-*ras* oncogene, which effectively neoplastically transformed NIH/3T3 and CHEF cells, induces no phenotypic changes in human cells. Doniger et al.[267] reported that fibroblasts from individuals with Bloom's

Table 2
COMPARISON OF TRANSFORMED PHENOTYPES
OF HUMAN VS. RODENT FIBROBLASTS INDUCED
BY CARCINOGENS

Trait	Rodent cells	Human cells
Morphological transformation	Common (10^{-2} to 10^{-3})	Rare (10^{-5} to 10^{-6})
Anchorage-independent growth	Rare ($\sim 10^{-6}$)	Common (10^{-2} to 10^{-3})
Immortality	Common	Rare
Progression to tumorigenic state	Common	Rare

syndrome following transfection with v-Ha-*ras* gene have an extended life span, but are not immortal, form colonies in agarose, and proliferate to a limited extent (but not progressively) in nude mice. Sutherland[268] has transfected human fibroblasts with DNA from malignant melanomas producing cells that vary in their ability to form foci and grow in soft agar, but are nontumorigenic. Human fibroblasts transfected with SV-40 DNA are morphologically altered and produce the viral tumor antigen, but do not form tumors in nude mice.[269] Rhim et al.[270] have shown that primary cultures of human epidermal keratinocytes infected with adeno 12-SV40 hybrid DNA viruses become immortal, but are not tumorigenic. However, if these cells are subsequently infected with Kirsten murine sarcoma virus (containing the Ki-*ras* oncogene), the cells form carcinomas when transplanted into nude mice. This important observation suggests that the two-step paradigm of immortalization and transformation has some validity in a human cell model. However, the extreme difficulty in neoplastically transforming human cells indicates that major differences exist in the human and rodent cell transformation models. In order to understand these differences, it is informative to compare the induction of various transformed phenotypes in human vs. rodent cells (Table 2). The differences observed may provide clues to understanding human carcinogenesis.

B. Induction of Morphological Transformation

Carcinogen treatment of certain rodent cells, i.e., Syrian hamster embryo cells, results in induction of morphologically altered colonies as discussed earlier. These cells have an increased propensity to develop into immortal cell lines ultimately capable of forming tumors in nude mice. Numerous studies utilizing either normal human diploid fibroblasts[253,254,263,265] or genetically deficient fibroblasts[258,271] report induction of a qualitatively similar morphological change in cultured human cells; however, morphologically transformed colonies comprised of human cells are less distinct than those of rodent cells and occasionally revert to normal morphology.[255,258] These results with human cells complicate attempts to isolate and quantitate morphologically transformed colonies. Kakunaga[253] estimated the frequency of morphological transformation in normal human fibroblasts treated with 4-NQO as 1 to 3 × 10^{-7} which is three to four orders of magnitude less than the frequency of morphological transformation with Syrian hamster embryo cells treated with carcinogens.[61]

C. Induction of Anchorage Independence

As discussed earlier in this chapter, induction of the anchorage-independent phenotype in cultured rodent cells is generally observed in immortal cell lines and is a rare, late-stage event in the progression towards malignancy. In contrast, anchorage-independent growth of normal diploid human fibroblasts is often easily induced by a variety of carcinogenic treatments. Chemical and physical carcinogens induce heritably stable anchorage-independent growth.[255-264] Silinskas et al.[260] were able to induce anchorage-independent colony formation at a high frequency (2.6×10^{-2}) by treating synchronous colonies of normal diploid human fibroblasts with propane sultone shortly after the onset of S phase. Furthermore, they were able to dramatically increase the frequency of colony growth in soft agar by propagating colonies initially isolated from soft agar and reassaying them in soft agar, suggesting that anchorage independence is heritable. McCormick et al.[272] reported a frequency of 1 to 3×10^{-2} for anchorage-independent human cells following treatment with MNNG. Although lower frequencies are also reported,[257] it is clear that this phenotypic change is readily induced by carcinogens in diploid human cells. These cells, however, are not tumorigenic or immortal.[257]

Culture conditions also have been demonstrated to exert an important effect on expression of the anchorage-independent phenotype of human fibroblast cells. Induction of anchorage-independent growth of normal human fibroblasts is achieved by utilizing high serum concentrations[273] or by addition of fibroblast growth factor or PDGF into the assay medium.[274] However, this expression of the anchorage-independent phenotype induced by culture conditions is reversible in contrast to carcinogen-induced anchorage-independent cells.[257] Growth in soft agar is also irreversibly altered by transfection with oncogenes.[267,268] Since oncogenes are often involved in growth factor responses,[59] these observations may be related.

D. Induction of Immortality

As discussed previously immortality is an important step in the neoplastic progression of rodent cells in culture. Escape from cellular senescence occurs spontaneously with a high frequency in some species, i.e., the mouse, and is readily induced in other rodent cells. In contrast, induction of immortal human fibroblasts is very rare.[140-143,253,257] Interestingly, induction of immortal human cell lines of nonmesenchymal origin have been reported recently. Stampfer and Bartley[139] observed rare, immortal cell lines in BaP-exposed mammary epithelial cell cultures. Yoakum et al.[105] and Harris et al.[106] obtained immortal human bronchial cells after transfection with the ras oncogene. Lechner et al.[275] were able to establish apparently immortal human mesothelial cells after treatment with asbestos. Immortalization of human epithelial cells by SV-40 virus is also more frequent than immortalization of human fibroblasts.[145] In each of these cases immortal cell lines are still very rare. However, considering the great extent of work done with fibroblasts and the lack of establishment of immortal cell lines with these cells, it is possible that nonfibroblast-like human cells may be easier to establish after carcinogen or oncogene treatment. Further quantitative studies in this area are needed. These differences may, of course, only be due to technical reasons (i.e., culture conditions used may be better for epithelial cell lines).

As discussed in more detail earlier, acquisition of immortality appears to be a multistep process. In the case of the BaP-treated human mammary epithelial cells[139] or SV-40-treated human fibroblasts,[140-143] cultures with an extended life span are often observed and rarely an immortal cell line develops out of these cultures. Furthermore, immortality appears to be recessive in cell hybrids between normal and immortal tumor cells.[146-151] For example, Bunn and Tarrant[147] fused Lesch-Nyhan cells at approximately 50% of their life span (population doubling 17) with HeLa cells, generating hybrids which survive only slightly longer than the parental cells. However, in 3 of 38 hybrid clonal cultures, foci of rapidly dividing cells appeared which eventually overgrew the entire culture. Morphologically transformed cells isolated from these foci and maintained in culture for 40 to 100 population doublings show no signs of senescence.

Escape from senescence was estimated to occur at a frequency of approximately 5 × 10^{-6} per hybrid cell. In another study involving fusion of normal human diploid fibroblasts with various immortal cell lines, rare variant immortal cells occurred at very low frequencies of about one in 10^{-5} cells.[146]

Thus, induction of the immortal phenotype, which is presumably a necessary or advantageous step in the progression towards neoplastic transformation, is a recessive event in cultured human cells. The rare occurrence of immortal human cells is distinctly different from the results with rodent cells and may be an important species difference.

E. Possible Differences Between Human and Rodent Cell Transformation

Several possible explanations can account for the difficulty to neoplastically transform normal human cells by treatment with carcinogens.

Culture conditions utilized in cell transformation studies were developed for use with rodent cells and may, therefore, be inappropriate or suboptimal for use with human cells. Other studies have also shown that assay conditions exert a profound influence over the expression of anchorage-independent growth[257,260,273,274] and life span.[276] Therefore, it is possible that the culture conditions developed to optimize growth of rodent cells may actually inhibit the expression of some feature of the malignant phenotype in human cells.

As discussed above, one of the key differences between rodent and human cells is the occurrence of immortal cell lines. This is frequent in all rodent cells (either spontaneously or after carcinogen treatment) and is rare in human cells even after multiple carcinogen treatments. The basis for this difference is unknown. One possibility is the difference in karyotypic stability of human vs. rodent cells. Aneuploidy has been postulated as a key change in the induction of immortality in rodent cells.[128,129] The need for multiple changes resulting from chromosomal changes in a karyotypically stable population may account for the lack of isolation of immortal and/or neoplastic human cells in culture.

Another possibility is that human cells require more steps for neoplastic transformation than rodent cells, and, therefore, this process will occur less frequently in human cells. One possibility is that additional steps are needed to activate human oncogenes. For example, Blair et al.[277] have shown that the mouse c-*mos* oncogene can be activated by a viral promoter (LTR) whereas this is not sufficient for activation of the human c-*mos* oncogene. An additional change is needed for activation of this gene. Thus, the number of steps may be different for mouse vs. human oncogenesis due to additional mechanisms controlling the inappropriate activation of oncogenes. Such control systems may have been necessary for the evolution of species with longer life spans.

An alternative pathway for a species to evolve control mechanisms to limit the inadvertant activation of oncogenes involves tumor suppression genes. Although the existence of such genes in human and rodent cells is now recognized, the number of these genes and differences of these genes in different species are unknown (see Chapter 3, Volume I for more details).

VII. CONCLUSION

There is clear evidence for a multistep process in the neoplastic transformation of normal cells in culture. This conclusion arises from studies of carcinogen-induced transformation and oncogene-mediated transformation. Two key steps are evident from both types of studies — an early change in the life span of the population resulting in escape from cellular senescence and immortality and a later change resulting in neoplastic transformation of these immortal cells. Evidence also exists that this two-step model is not sufficient to explain all the results. Immortality appears to be a recessive trait, and multiple steps are required for cells to acquire this phenotypic change. Also, multiple steps are sometimes required for immortal cells to progress to neoplastic potential. One key additional step is the loss of the ability of normal cells to suppress tumorigenicity of malignant cells in cell hybrids. These findings implicate in the neoplastic process another class of genes, tumor suppressor genes, in addition to oncogenes. These tumor suppressor genes have to be lost or inactivated for neoplastic transformation to occur. It appears that neoplastic transformation of human cells is distinctly less frequent than with rodent cells. The basis for this difference is unknown, but requires additional exploration before an understanding of the key changes in carcinogenesis can be achieved.

REFERENCES

1. Foulds, L., *Neoplastic Development,* Vol. 1, Academic Press, London, 1969.
2. Boutwell, R. K., The function and mechanism of promoters of carcinogenesis, *CRC Crit. Rev. Toxicol.,* 2, 419, 1974.
3. Hennings, H., Tumor promotion and progression in mouse skin, in *Mechanisms of Environmental Carcinogenesis,* Vol. 2, Barrett, J. C., Ed., CRC Press, Boca Raton, Fla., 1987, chap. 11.
4. Knudson, A. G., Mutation and human cancer, *Adv. Cancer Res.,* 17, 317, 1973.
5. Hethcote, H. W. and Knudson, A. G., Model for the incidence of embryonal cancers: application to retinoblastoma, *Proc. Natl. Acad. Sci. U.S.A.,* 75, 2453, 1978.
6. Ashley, D. J. B., The two "hit" and multiple "hit" theories of carcinogenesis, *Br. J. Cancer,* 23, 313, 1969.
7. Knudson, A. G., Hethcote, H. W., and Brown, B. W., Mutation and childhood cancer: a probabilistic model for the incidence of retinoblastoma, *Proc. Natl. Acad. Sci. U.S.A.,* 72, 5116, 1975.
8. Cavenee, W. K., Dryja, T. P., Phillips, R. A., Benedict, W. F., Godbout, R., Gallie, B. L., Murphree, A. L., Strong, L. C., and White, R. L., Expression of recessive alleles by chromosomal mechanisms in retinoblastoma, *Nature (London),* 305, 779, 1983.
9. Kaldor, J. M. and Day, N. E., Interpretation of epidemiological studies in the context of the multistage model of carcinogenesis, in *Mechanisms of Environmental Carcinogenesis,* Vol. 2, Barrett, J. C., Ed., CRC Press, Boca Raton, Fla., 1987, chap. 10.
10. Doll, R., An epidemiological perspective of the biology of cancer, *Cancer Res.,* 38, 3573, 1978.
11. Doll, R. and Peto, R., The causes of cancer, *J. Natl. Cancer Inst.,* 66, 1197, 1981.
12. Barrett, J. C. and Ts'o, P. O. P., Evidence for the progressive nature of neoplastic transformation in vitro, *Proc. Natl. Acad. Sci. U.S.A.,* 75, 3761, 1978.
13. Mondal, S., Brankow, D. W., and Heidelberger, C., Two-stage chemical oncogenesis in culture of C3H/10T1/2 cells, *Cancer Res.,* 36, 2254, 1976.
14. Kennedy, A. R., Promotion and other interactions between agents in the induction of transformation in vitro in fibroblast-like cell culture systems, in *Mechanisms of Tumor Promotion,* Vol. 3, Slaga, T. J., Ed., CRC Press, Boca Raton, Fla., 1984, 13.
15. Barrett, J. C. and Thomassen, D. G., Use of quantitative cell transformation assays in risk estimation, in *Methods of Estimating Risk in Human and Chemical Damage in Non-human Biota and Ecosystems, SCOPE, SGOMSEC 2, IPCS Joint Symposia 3,* Vouk, V. B., Butler, G. C., Hoel, D. G., and Peakall, D. B., Eds., John Wiley & Sons, New York, 1985, 201.
16. Barrett, J. C., Cell culture models of multistep carcinogenesis, in *IARC Scientific Publications No. 58, Age-Related Factors in Carcinogenesis,* Likhachev, A., Anisimov, V., and Montesano, R., Eds., International Agency for Research on Cancer, Lyon, 1985, 181.
17. Land, H., Parada, L. F., and Weinberg, R. A., Tumorigenic conversion of primary embryo fibroblasts requires at least two cooperating oncogenes, *Nature (London),* 304, 596, 1983.
18. Ruley, H. E., Adenovirus early region 1A enables viral and cellular transforming genes to transform primary cells in culture, *Nature (London),* 304, 602, 1983.
19. Newbold, R. F. and Overell, R. W., Fibroblast immortality is a prerequisite for transformation by EJ c-Ha-*ras* oncogene, *Nature (London),* 304, 648, 1983.

20. Rassoulzadegan, M., Naghashfar, Z., Cowie, A., Carr, A., Grisoni, M., Kamen, R., and Cuzin, F., Expression of the large T protein of polyoma virus promotes the establishment in culture of "normal" rodent fibroblast cell lines, *Proc. Natl. Acad. Sci. U.S.A.*, 80, 4354, 1983.
21. Thomassen, D. G., Gilmer, T. G., Annab, L. A., and Barrett, J. C., Evidence for multiple steps in neoplastic transformation of normal and preneoplastic Syrian hamster embryo cells following transfection with Harvey murine sarcoma virus oncogene (v-Ha-*ras*), *Cancer Res.*, 45, 726, 1985.
22. Oshimura, M., Gilmer, T. M., and Barrett, J. C., Nonrandom loss of chromosome 15 in Syrian hamster tumors induced by v-Ha-*ras* plus v-*myc* oncogenes, *Nature (London)*, 316, 636, 1985.
23. Barrett, J. C., A multistep model for neoplastic development: role of genetic and epigenetic changes, in *Mechanisms of Environmental Carcinogenesis*, Vol. 2, Barrett, J. C., Ed., CRC Press, Boca Raton, Fla., 1987, chap. 13.
24. Berwald, Y. and Sachs, L., *In vitro* cell transformation with chemical carcinogens, *Nature (London)*, 200, 1182, 1963.
25. Berwald, Y. and Sachs, L., *In vitro* transformation of normal cells to tumor cells by carcinogenic hydrocarbons, *J. Natl. Cancer Inst.*, 35, 641, 1965.
26. Hayflick, L., Recent advances in the cell biology of aging, *Mech. Aging Dev.*, 14, 59, 1981.
27. Newbold, R. F., Overell, R. W., and Connell, J. R., Induction of immortality is an early event in malignant transformation of mammalian cells by carcinogens, *Nature (London)*, 299, 633, 1982.
28. Benedict, W. F., Rucker, N., Mark, C., and Kouri, R. E., Correlation between balance of specific chromosomes and expression of malignancy in hamster cells, *J. Natl. Cancer Inst.*, 54, 157, 1975.
29. Huberman, E. and Sachs, L., Cell susceptibility to transformation and cytotoxicity by the carcinogenic hydrocarbon benzo(a)pyrene, *Proc. Natl. Acad. Sci. U.S.A.*, 56, 1123, 1966.
30. Barrett, J. C., Crawford, B. D., Mixter, L. O., Schechtman, T. L., Ts'o, P. O. P., and Pollack, R., Correlation of *in vitro* growth properties and tumorigenicity of Syrian hamster cell lines, *Cancer Res.*, 39, 1504, 1979.
31. DiPaolo, J. A., Donovan, P. J., and Nelson, R. L., *In vitro* transformation of hamster cells by polycyclic hydrocarbons: factors influencing the number of cells transformed, *Nature (London) New Biol.*, 230, 240, 1971.
32. DiPaolo, J. A., Donovan, J., and Nelson, R. L., Transformation of hamster cells *in vitro* by polycyclic hydrocarbons without cytotoxicity, *Proc. Natl. Acad. Sci. U.S.A.*, 68, 2958, 1971.
33. DiPaolo, J. A., Nelson, R. L., and Donovan, P. J., Morphological, oncogenic, and karyological characteristics of Syrian hamster embryo cells transformed *in vitro* by carcinogenic polycyclic hydrocarbons, *Cancer Res.*, 31, 1118, 1971.
34. Borek, C. and Sachs, L., *In vitro* cell transformation by X-irradiation, *Nature (London)*, 210, 276, 1966.
35. DiPaolo, J. A. and Donovan, P. J., Properties of Syrian hamster cells transformed in the presence of carcinogenic hydrocarbons, *Exp. Cell Res.*, 48, 1361, 1967.
36. Kuroki, T. and Sato, H., Transformation and neoplastic development *in vitro* of hamster embryonic cells by 4-nitroquinoline-1-oxide and its derivatives, *J. Natl. Cancer Inst.*, 41, 53, 1968.
37. Kamahora, J. and Kakunaga, T., *In vitro* carcinogenesis of 4-ntiroquinoline-1-oxide with golden hamster embryonic cells, *Proc. Jpn. Acad.*, 42, 1979, 1966.
38. Kakunaga, T. and Kamahora, J., Analytical studies on the process of malignant transformation of hamster cells in cultures with 4-nitroquinoline-1-oxide, *Symp. Cell. Chem.*, 20, 135, 1969.
39. Barrett, J. C., A preneoplastic stage in the spontaneous neoplastic transformation of Syrian hamster embryo cells in culture, *Cancer Res.*, 40, 91, 1980.
40. Gilmer, T. M., Annab, L. A., Oshimura, M., and Barrett, J. C., Neoplastic transformation of normal and carcinogen-induced preneoplastic Syrian hamster embryo cells by the v-*src* oncogene, *Mol. Cell Biol.*, 5, 1707, 1985.
41. Evans, C. H. and DiPaolo, J. A., Independent expression of chemical carcinogen-induced phenotypic porperties frequently associated with the neoplastic state in a cultured guinea pig cell line, *J. Natl. Cancer Inst.*, 69, 1175, 1982.
42. Evans, C. H. and DiPaolo, J. A., Neoplastic transformation of guinea pig fetal cells in culture induced by chemical carcinogens, *Cancer Res.*, 35, 1035, 1975.
43. Ponten, J., Spontaneous and virus induced transformation in cell culture, *Virol. Monogr.*, 8, 1, 1971.
44. Nettleship, A. and Earle, W. R., Production of malignancy *in vitro*. VI. Pathology of tumors produced, *J. Natl. Cancer Inst.*, 4, 229, 1943.
45. Aaronson, S. A. and Todaro, G. J., Basis for the acquisition of malignant potential by mouse cells cultivated *in vitro*, *Science*, 162, 1024, 1968.
46. Nettesheim, P. and Barrett, J. C., Tracheal epithelial cell transformation: a model system for studies on neoplastic progression, *CRC Crit. Rev. Toxicol.*, 12, 215, 1984.

47. Colburn, N. H., Bruegge, W. F. V., Bates, J. R., Gray, R. H., Roseen, J. D., Kelsey, W. H., and Shimada, T., Correlation of anchorage-independent growth with tumorigenicity of chemically transformed mouse epidermal cells, *Cancer Res.*, 38, 624, 1978.
48. Yuspa, S. H., Hawley-Nelson, P., Koehler, B., and Stanley, J. R., A survey of transformation markers in differentiating epidermal cell lines in culture, *Cancer Res.*, 40, 4694, 1980.
49. Bishop, J. M., Viral oncogenes, *Cell*, 42, 23, 1985.
50. Bishop, J. M., Cellular oncogenes and retroviruses, *Annu. Rev. Biochem.*, 52, 301, 1983.
51. Shin, C., Shilo, B. Z., Goldfarb, M. P., Dannenberg, A., and Weinberg, R. A., Passage of phenotypes of chemically transformed cells via transfection of DNA and chromatin, *Proc. Natl. Acad. Sci. U.S.A.*, 76, 5714, 1979.
52. Cooper, G. M., Okenquist, S., and Silverman, L., Transforming activity of DNA of chemically transformed and normal cells, *Nature (London)*, 284, 418, 1980.
53. Perucho, M., Goldfarb, M., Shimizu, K., Lama, C., Fogh, J., and Wigler, M., Human-tumor-derived cell lines contain common and different transforming genes, *Cell*, 27, 467, 1981.
54. Weinberg, R. A., Oncogenes of spontaneous and chemically induced tumors, *Adv. Cancer Res.*, 36, 149, 1982.
55. Varmus, H. E., The molecular genetics of cellular oncogenes, *Annu. Rev. Genet.*, 18, 553, 1984.
56. Mougneau, E., Lemieux, L., Rassoulzadegan, M., and Cuzin, F., Biological activities of v-*myc* and rearranged c-*myc* oncogenes in rat fibroblast cells in culture, *Proc. Natl. Acad. Sci. U.S.A.*, 81, 5758, 1984.
57. Gionti, E., Pontarelli, G., and Cancedda, R., Avain myelocytomatosis virus immortalizes differentiated quail chondrocytes, *Proc. Natl. Acad. Sci. U.S.A.*, 82, 2756, 1985.
58. Land, H., Parda, L. F., and Weinberg, R. A., Cellular oncogenes and multistep carcinogenesis, *Science*, 222, 771, 1983.
59. Weinberg, R. A., The action of oncogenes in the cytoplasm and nucleus, *Science*, 230, 770, 1985.
60. Barrett, J. C., Hesterberg, T. W., and Thomassen, D. G., Use of cell transformation systems for carcinogenicity testing and mechanistic studies of carcinogenesis, *Pharmacol. Rev.*, 36, 53S, 1984.
61. Barrett, J. C. and Ts'o, P. O. P., Relationship between somatic mutation and neoplastic transformation, *Proc. Natl. Acad. Sci. U.S.A.*, 75, 3297, 1978.
62. Barrett, J. C. and Lamb, P. W., Tests with the Syrian hamster embryo cell transformation assay, in *Progress in Mutation Research*, Vol. 5, Ashby, J., deSerres, F. J., Draper, M., Ishidte, M., Margolin, B. H., Matter, B. E., and Shelby, M. D., Eds., Elsevier, Amsterdam, 1985, 623.
63. Gart, J. J., DePaolo, J. A., and Donovan, P. J., Mathematical models and the statistical analyses of cell transformation experiments, *Cancer Res.*, 39, 6069, 1979.
64. Huberman, E., Mager, R., and Sachs, L., Mutagenesis and transformation of normal cells by chemical carcinogens, *Nature (London)*, 264, 360, 1976.
65. Siliciano, M. J., Adair, G. M., Atkinson, E. N., and Humphrey, R. M., Induced somatic cell mutations detected in cultured cells by electrophoresis, *Isozymes Curr. Top. Biol. Med. Res.*, 10, 41, 1983.
66. Steglich, C. S. and DeMars, R., Mutations causing deficiency of APRT in fibroblasts cultured from humans heterozygous for mutant APRT alleles, *Somatic Cell Genet.*, 8, 115, 1982.
67. Barrett, J. C., Cell transformation, mutation, and cancer, *Gann Monogr. Cancer Res.*, 27, 195, 1981.
68. Barrett, J. C., Tsutsui, T., and Ts'o, P. O. P., Neoplastic transformation induced by a direct perturbation of DNA, *Nature (London)*, 274, 229, 1978.
69. Tsutsui, T., Barrett, J. C., and Ts'o, P. O. P., Morphological transformation, DNA damage, and chromosome aberrations induced by a direct DNA pertubation of synchronized Syrian hamster embryo cells, *Cancer Res.*, 39, 2356, 1979.
70. Lin, S. L., Takii, M., and Ts'o, P. O. P., Somatic mutation and neoplastic transformation induced by [methyl-^{3}H] thymidine, *Radiat. Res.*, 90, 142, 1982.
71. Barrett, J. C., Gene mutation and cell transformation of mammalian cells induced by two modified purines: 2-aminopurine and 6-N-hydroxylaminopurine, *Proc. Natl. Acad. Sci. U.S.A.*, 78, 5685, 1981.
72. Zajac-Kaye, M. and Ts'o, P. O. P., DNase I encapsulated in liposomes can induce neoplastic transformation of Syrian hamster embryo cells in culture, *Cell*, 39, 427, 1984.
73. Barrett, J. C., Hesterberg, T. W., Oshimura, M., and Tsutsui, T., Role of chemically induced mutagenic events in neoplastic transformation of Syrian hamster embryo cells, in *Carcinogenesis — A Comprehensive Survey*, Vol. 9, Barrett, J. C. and Tennant, R. W., Eds., Raven Press, New York, 1985, 123.
74. Doniger, J., Jacobson, E. D., Krell, K., and DiPaolo, J. A., Ultraviolet light action spectra for neoplastic transformation and lethality of Syrian hamster embryo cells correlate with spectrum for pyrimidine dimer formation in cellular DNA, *Proc. Natl. Acad. Sci. U.S.A.*, 78, 2378, 1981.
75. Doniger, J. and DiPaolo, J. A., Modulation of *in vitro* transformation and the early and late modes of DNA replication of UV-irradiated Syrian hamster cells by caffeine, *Radiat. Res.*, 87, 568, 1981.

76. Barrett, J. C., Relationship between mutagenesis and carcinogenesis, in *Mechanisms of Environmental Carcinogenesis*, Vol. 1, Barrett, J. C., Ed., CRC Press, Boca Raton, Fla., 1987, chap. 8.
77. deKok, A. J., Tates, A. D., Ben Engdse, L., and Simons, J. W. I. M., Genetic and molecular mechanisms of the *in vitro* transformation of Syrian hamster embryo cells by the carcinogen N-ethyl-N-nitrosourea. I. Correlation of morphological transformation and enhanced fibrinolytic activity to gene mutation, chromosomal alterations and lethality, *Carcinogenesis*, 6, 1565, 1985.
78. Gilmer, T. M., Annab, L. A., and Barrett, J. C., unpublished data.
79. Nakano, S. and Ts'o, P. O. P., Cellular differentiation and neoplasia: characterization of subpopulations of cells that have neoplasia-related growth properties in Syrian hamster embryo cell cultures, *Proc. Natl. Acad. Sci. U.S.A.*, 78, 4995, 1981.
80. Nakano, S., Bruce, S. A., Ueo, H., and Ts'o, P. O. P., A qualitative and quantitative assay for cells lacking postconfluence inhibition of cell division: characterization of this phenotype in carcinogen-treated Syrian hamster embryo cells in culture, *Cancer Res.*, 42, 3132, 1982.
81. Nakano, S., Bruce, S. A., Ueo, H., and Ts'o, P. O. P., A contact-insensitive subpopulation in Syrian hamster cell cultures with a greater susceptibility to chemically induced neoplastic transformation, *Proc. Natl. Acad. Sci. U.S.A.*, 82, 5005, 1985.
82. Ts'o, P. O. P., Bruce, S. A., Brown, A. R., and Miller, P. S., New view of carcinogenesis: implication for chemotherapy and human risk assessment, in *Carcinogenesis — A Comprehensive Survey*, Vol. 9, Barrett, J. C. and Tennant, R. W., Eds., Raven Press, New York, 1985, 105.
83. Nettesheim, P. and Barrett, J. C., *In vitro* transformation of rat tracheal epithelial cells as a model for the study of multistage carcinogenesis, in *Carcinogenesis — A Comprehensive Survey*, Vol. 9, Barrett, J. C. and Tennant, R. W., Eds., Raven Press, New York, 1985, 283.
84. Nettesheim, P. and Marchok, A., Neoplastic development in airway epithelium, *Adv. Cancer Res.*, 39, 1, 1983.
85. Marchok, A. C., Nettsheim, P., and Johnston, W. W., Localization of specific lesions in dimethylbenz(a)anthracene-pre-exposed tracheal explants, *Am. Assoc. Pathol.*, 109, 321, 1982.
86. Terzaghi, M. and Nettesheim, P., Dynamics of neoplastic development in carcinogen-exposed tracheal mucosa, *Cancer Res.*, 39, 4003, 1979.
87. Terzaghi, M., Nettesheim, P., and Riester, L., Effect of carcinogen dose on the dynamics of neoplastic development in rat tracheal epithelium, *Cancer Res.*, 42, 4511, 1982.
88. Thomassen, D. G., Gray, T., Mass, M. J., and Barrett, J. C., High frequency of carcinogen-induced early, preneoplastic changes in rat tracheal epithelial cells in culture, *Cancer Res.*, 43, 5956, 1983.
89. Yuspa, S. H., Mechanisms of transformation and promotion of mouse epidermal cells, in *Carcinogenesis — A Comprehensive Survey*, Vol. 9, Barrett, J. C. and Tennant, R. W., Eds., Raven Press, New York, 1985, 283.
90. Hennings, H., Michael, D., Cheng, C., Steinert, P., Holbrook, K., and Yuspa, S. H., Calcium regulation of growth and differentiation of mouse epidermal cells in culture, *Cell*, 19, 245, 1980.
91. Kulesz-Martin, M. F., Koehler, B., Hennings, H., and Yuspa, S. H., Quantitative assay for carcinogen altered differentiation in mouse epidermal cells, *Carcinogenesis*, 1, 995, 1980.
92. Yuspa, S. H. and Morgan, D. L., Mouse skin cells resistant to terminal differentiation associated with initiation of carcinogenesis, *Nature (London)*, 293, 72, 1981.
93. Kulesz-Martin, M., Kilkenny, A. E., Holbrook, K. A., Digernes, V., and Yuspa, S. H., Properties of carcinogen altered mouse epidermal cells resistant to calcium-induced terminal differentiation, *Carcinogenesis*, 4, 1367, 1983.
94. Kilkenny, A. E., Morgan, D., Spangler, E. F., and Yuspa, S. H., Correlation of initiating potency of skin carcinogens with potency to induce resistance to terminal differentiation in cultured mouse keratinocytes, *Cancer Res.*, 45, 2219, 1985.
95. Toftgard, R., Yuspa, S. H., and Roop, D. R., Keratin gene expression in mouse skin tumors and in mouse skin treated with 12-O-tetradecanoyl-phorbol-13-acetate, *Cancer Res.*, 45, 5845, 1985.
96. Balmain, A. and Pragnell, I. B., Mouse skin carcinomas induced *in vivo* by chemical carcinogens have a transforming Harvey-*ras* oncogene, *Nature (London)*, 303, 72, 1983.
97. Balmain, A., Ramsden, M., Bouden, G. T., and Smith, J., Activation of the mouse cellular Harvey-*ras* gene in chemically induced benign skin papillomas, *Nature (London)*, 307, 658, 1984.
98. Balmain, A., Transforming *ras* oncogenes and multistage carcinogenesis, *Br. J. Cancer*, 51, 1, 1985.
99. Toftgard, R., Roop, D. R., and Yuspa, S. H., Proto-oncogene expression during two-stage carcinogenesis in mouse skin, *Carcinogenesis*, 6, 655, 1985.
100. Balmain, A., personal communication.
101. Yuspa, S. H., Vass, W., and Scolnick, E., Altered growth and differentiation of cultured mouse epidermal cells infected with oncogenic retrovirus: contrasting effects of viruses and chemicals, *Cancer Res.*, 43, 6021, 1983.
102. Weismann, B. E. and Aaronson, S. A., BALB and Kirsten murine sarcoma viruses alter growth and differentiation of EGF-dependent BALB/c mouse epidermal keratinocyte lines, *Cell*, 32, 599, 1983.

103. Jetten, A. M., Fitzgerald, D. J., and Nettesheim, P., Control of differentiation and proliferation of normal and transformed airway epithelial cells by retinoids, in *Nutritional Diseases: Research Directions in Comparative Pathobiology,* Miyaki, G., Ed., Alan R. Liss, New York, 1986.
104. Kulesz-Martin, M., Yoshida, M. A., Prestine, L., Yuspa, S. H., and Bertram, J. S., Mouse cell clones for improved quantitation of carcinogen-induced altered differentiation, *Carcinogenesis,* 6, 1245, 1985.
105. Yoakum, G. H., Lechner, J. F., Gabrielson, E., Korba, B. E., Malan-Shilbey, L., Willey, J. C., Valerio, M. G., Shamsuddin, A. M., Trump, B. F., and Harris, C. C., Transformation of human bronchial epithelial cells transfected by Harvey-*ras* oncogene, *Science,* 227, 1174, 1985.
106. Harris, C. C., Lechner, J. F., Yoakum, G. H., Amstad, P., Korba, B. E., Gabrielson, E., Grafstrom, R., Shamsuddin, A., and Trump, B. V., *In vitro* studies of human lung carcinogenesis, in *Carcinogenesis — A Comprehensive Survey,* Vol. 9, Barrett, J. C. and Tennant, R. W., Eds., Raven Press, New York, 1985, 257.
107. Hayflick, L., Cell death *in vitro* in *Cell Death in Biology and Pathology,* Bowen, I. D. and Lockshin, R. A., Eds., Chapman and Hall, New York, 1981, 243.
108. Todaro, G. J. and Green, H., Serum albumin supplemented medium for long term cultivation of mammalian fibroblast strains, *Proc. Soc. Exp. Biol. Med.,* 116, 668, 1964.
109. Todaro, G. J., Nikausen, K., and Green, H., Growth properties of polyoma virus-induced hamster tumor cells, *Cancer Res.,* 23, 825, 1963.
110. Rheinwald, J. G. and Green, H., Epidermal growth factor and the multiplication of cultured human epidermal keratinocytes, *Nature (London),* 265, 421, 1977.
111. Todaro, G. J. and Green, H., Quantitative studies of the growth of mouse embryo cells in culture and their development into established lines, *J. Cell Biol.,* 17, 229, 1963.
112. Matsuya, Y. and Yamane, I., Serial culture of Syrian hmaster fibroblasts in albumin fortified medium and their regular development into established lines, *Exp. Cell Res.,* 50, 652, 1968.
113. Thomassen, D. G., Role of spontaneous transformation in carcinogenesis: development of preneoplastic rat tracheal epithelial cells at a constant rate, *Cancer Res.,* 46, 2344, 1986.
114. Rassoulzadegan, M., Cowie, A., Carr, A., Glaichenhause, N., Kamen, R., and Cuzin, F., The roles of individual polyoma virus early proteins in oncogenic transformation, *Nature (London),* 300, 713, 1982.
115. Houweling, A., van den Elsen, P. J., and van der Eb, A. J., Partial transformation of primary rat cells by the leftmost 4.5% fragment of adenovirus 5 DNA, *Virology,* 105, 537, 1980.
116. van den Elsen, P., de Pater, S., Houweling, A., van der Veer, J., and van der Eb, A., The relationship between region E1a and E1b of human adenoviruses in cell transformation, *Gene,* 18, 175, 1982.
117. Griffin, B. E. and Karran, L., Immortalization of monkey epithelial cells by specific fragments of Epstein-Barr virus DNA, *Nature (London),* 309, 78, 1984.
118. Jenkins, J. R., Rudge, K., and Currie, G. A., Cellular immortalization by a cDNA clone encoding the transformation-associated phosphoprotein *p53, Nature (London),* 312, 651, 1984.
119. Parada, L. F., Lanol, H., Weinberg, R. A., Wolf, D., and Rotter, V., Cooperation between gene encoding *p53* tumour antigen and *ras* in cellular transformation, *Nature (London),* 312, 649, 1984.
120. Eliyahu, D., Michalovitz, D., and Oren, M., Overproduction of *p53* antigen makes established cells highly tumorigenic, *Nature (London),* 316, 158, 1985.
121. Kelly, K., Cochran, B. H., Stiles, C. D., and Leder, R., Cell-specific regulation of the c-*myc* gene by lymphocyte mitogens and platelet-derived growth factor, *Cell,* 35, 603, 1983.
122. Armelin, H. A., Armelin, M. C. S., Kelly, K., Stewart, T., Leder, P., Cochran, B. H., and Stiles, C. D., Functional role for c-*myc* in mitogenic response to platelet-derived growth factor, *Nature (London),* 310, 655, 1984.
123. Kaczmarek, L., Hyland, J. K., Watt, R., Rosenberg, M., and Baserga, R., Microinjected c-*myc* as a competence factor, *Science,* 14, 1313, 1985.
124. Rapp, U. R., Cleveland, J. L., Brightman, K., Scott, A., and Ihle, J. N., Abrogation of Il-3 and Il-2 dependence by recombinant murine retroviruses expressing v-*myc* oncogene, *Nature (London),* 317, 434, 1984.
125. Balk, S. D., Riley, T. M., Gunther, H. S., and Morisi, A., Heparin-treated, v-*myc*-transformed chicken heart mesenchymal cells assume a normal morphology but are hypersensitive to epidermal growth factor (EGF) and brain fibroblast growth factor (bFGF); cells transformed by the v-Ha-*ras* oncogene are refractory to EGF and bFGF but are hypersensitive to insulin-like growth factors, *Proc. Natl. Acad. Sci. U.S.A.,* 82, 5781, 1985.
126. Gordon, J., Ley, S. C., Melamed, M. D., English, L. S., and Hughes-Jones, N. C., Immortalized B lymphocytes produce B-cell growth factor, *Nature (London),* 310, 145, 1984.
127. Palmieri, S., Kahn, P., and Graf, T., Quail embryo fibroblasts transformed by four v-*myc*-containing virus isolates show enhanced proliferation but are non tumorigenic, *EMBO J.,* 2, 2385, 1983.

128. Oshimura, M., Hesterberg, T. W., and Barrett, J. C., An early, non-random karyotypic change in immortal Syrian hamster cell lines transformed by asbestos: trisomy of chromosome 11, *Cancer Genet. Cytogenet.*, 22, 225, 1986.
129. Oshimura, M. and Barrett, J. C., Chemically induced aneuploidy in mammalian cells: mechanisms and biological significance in cancer, *Environ. Mutag.*, 8, 129, 1986.
130. Keath, E. J., Caimi, P. G., and Cole, M. D., Fibroblast lines expressing activated c-*myc* oncogenes are tumorigenic in nude mice and syngeneic animals, *Cell*, 39, 339, 1984.
131. Jenkins, J. R., Rudge, K., Chumakov, P., and Currie, G. A., The cellular oncogene *p53* can be activated by mutagenesis, *Nature (London)*, 317, 816, 1985.
132. Fitzgerald, D. J., Kitamura, H., Barrett, J. C., and Nettesheim, P., Analysis of growth fractions and stem cell compartments in transformed rat tracheal epithelial cell colonies, *Cancer Res.*, 46, 4642, 1986.
133. Thomassen, D. G., Barrett, J. C., Beeman, D. K., and Nettesheim, P., Changes in stem cell population of rat tracheal epithelial cell cultures at an early stage in neoplastic progression, *Cancer Res.*, 45, 3322, 1985.
134. Matsumura, T., Hayashi, M., and Konishi, R., Immortalization in culture of rat cells: a geneologic study, *J. Natl. Cancer Inst.*, 74, 1223, 1985.
135. Matsumura, T., Sequence of cell life phases in a finitely proliferative population of cultured rat cells: a geneologic study, *J. Cell. Physiol.*, 119, 145, 1984.
136. Matsumura, T., A rare family tree of cultured rat cells showing a change in proliferative potential, *Cell Biol. Int. Rep.*, 7, 931, 1983.
137. Matsumura, T., Masuda, K., Murakami, Y., and Konishi, R., Family trees representing the finitely proliferative nature of cultured rat liver cells, *Cell Struct. Function*, 8, 293, 1983.
138. Ethier, S. P., Primary culture and serial passage of normal and carcinogen-treated rat mammary epithelial cells *in vitro*, *J. Natl. Cancer Inst.*, 74, 1307, 1985.
139. Stampfer, M. R. and Bartley, J. C., Induction of transformation and continuous cell lines from normal human mammary epithelial cells after exposure to benzo(a)pyrene, *Proc. Natl. Acad. Sci. U.S.A.*, 82, 2394, 1985.
140. Huschtscha, L. I. and Holliday, R., Limited and unlimited growth of SV40-transformed cells from human diploid MRC-5 fibroblasts, *J. Cell Sci.*, 63, 77, 1983.
141. Girardi, A. J., Jensen, F. C., Koprowski, H., SV40-induced transformation of human diploid cells: crisis, and recovery, *J. Cell. Comp. Physiol.*, 65, 69, 1965.
142. Stein, G. H., SV40-transformed human fibroblasts: evidence for cellular aging in precrisis cells, *J. Cell. Physiol.*, 125, 36, 1985.
143. Moyer, A. W., Wallace, R., Cox, H. R., Limited growth period of human lung cell lines transformed by simian virus 40, *J. Natl. Cancer Inst.*, 33, 227, 1964.
144. Steinberg, M. L. and Defendi, V., Altered patterns of keratin synthesis in human epidermal keratinocytes transformed by SV40, *J. Cell. Physiol.*, 123, 117, 1985.
145. Defendi, V., Naimski, P., and Steinberg, M. L., Human cells transformed by SV40 revisited: the epithelial cells, *J. Cell. Physiol.*, 2, 131, 1982.
146. Pereira-Smith, O. M. and Smith, J. R., Evidence for the recessive nature of cellular immortality, *Science*, 221, 963, 1983.
147. Bunn, C. L. and Tarrant, G. M., Limited lifespan in somatic cell hybrids and cybrids, *Exp. Cell Res.*, 127, 385, 1980.
148. Stein, G. H., Namba, M., and Corsaro, C. M., Relationship of finite proliferative lifespan, senescence, and quiescence in human cells, *J. Cell. Physiol.*, 122, 343, 1985.
149. Pereira-Smith, O. M. and Smith, J. R., Expression of SV40 T antigen in finite lifespan hybrids of nomal and SV40-transformed fibroblasts, *Somatic Cell Genet.*, 7, 411, 1981.
150. Muggleton-Harris, A. L. and DeSimone, D. W., Replicative potentials of various fusion products between WI-38 and SV40-transformed WI-38 cells and their components, *Somatic Cell Genet.*, 6, 689, 1980.
151. Norwood, T. H., Pendergrass, W. R., Sprague, C. A., and Martin, G. M., Dominance of the senescent phenotype in heterokaryons between replicative and post-replicative fibroblast-like cells, *Proc. Natl. Acad. Sci. U.S.A.*, 71, 2231, 1974.
152. Daniel, C. W., deOme, K. B., Young, L. J. T., Blair, P. B., and Faulkin, J. J., Jr., The *in vivo* lifespan of normal and preneoplastic mouse mammary glands: a serial transplanation study, *Proc. Natl. Acad. Sci. U.S.A.*, 61, 53, 1968.
153. Krohn, P. L., Review lectures on senescence. II. Heterochronic transplantation in the study of aging, *Proc. R. Soc. London Ser. B*, 157, 128, 1962.
154. Ford, C. E., Michlem, H. S., and Gray, S. M., Evidence of selective proliferation of reticular cell-clones in heavily irradiated mice, *Br. J. Radiol.*, 32, 280, 1959.

155. Cudkowicz, G., Upton, A. C., Shearer, G. M., and Hughes, W. L., Lymphocyte content and proliferative capacity of serially transplanted mouse bone marrow, *Nature (London)*, 201, 165, 1964.
156. Siminovitch, L., Till, J. E., and McCulloch, E. A., Decline in colony-forming ability of marrow cells subjected to serial transplantation into irradiated mice, *J. Cell. Comp. Physiol.*, 64, 23, 1964.
157. Williamson, A. R. and Askonas, B. A., Senescence of an antibody-forming cell clone, *Nature (London)*, 238, 337, 1972.
158. Harrison, D. F., Normal function of transplanted marrow cell lines from aged mice, *J. Gerontol.*, 30, 279, 1975.
159. Hellman, S., Botnick, L. E., Hannon, E. C., and Vigneulle, R. M., Proliferative capacity of murine hematopoietic stem cells, *Proc. Natl. Acad. Sci. U.S.A.*, 75, 490, 1978.
160. Daniel, C. W. and Young, L. F. T., Influence of cell division on an aging process, *Exp. Cell Res.*, 65, 27, 1971.
161. Hayflick, L. and Moorhead, P. S., The serial cultivation of human diploid cell strains, *Exp. Cell Res.*, 25, 585, 1961.
162. Daniel, C. W., Aidells, B. D., Medina, D., and Faulkin, L. F., Jr., Unlimited division potential of precancerous mouse mammary cells after spontaneous or carcinogen-induced transformation, *Fed. Proc., Fed. Am. Soc. Exp. Biol.*, 34, 64, 1975.
163. Till, J. E., McCulloch, E. A., and Simonvitch, L., Isolation of variant cell lines during serial transplantation of hematopoietic cells derived from fetal liver, *J. Natl. Cancer Inst.*, 33, 707, 1964.
164. Stewart, H. L., Snell, K. C., Dunham, L. J., and Schylen, S. M., Transplantable and transmissible tumors of animals, Report, Armed Forces Institute of Pathology, Washington D.C., 1959.
165. Holliday, R., Cancer and cell senescence, *Nature (London)*, 306, 742, 1983.
166. Leibovitz, A., The establishment of cell lines from human solid tumors, *Adv. Cell Cult.*, 4, 249, 1985.
167. Bruland, O., Fostad, O., and Pihl, A., The use of multicellular spheroids in establishing human sarcoma cell lines *in vitro, Int. J. Cancer*, 35, 793, 1985.
168. Smith, H. S., Wolman, S. R., Aver, G., and Hackett, A. J., Cell culture studies: a perspective on malignant progressions of human breast cancer, in *Breast Cancer: On the Frontier of Discovery*, Rich, M. and Taylor, J., Eds., in press.
169. Smith, H. S., Liotta, L. A., Hancock, M. C., Wolman, S. R., and Hackett, A. J., Invasiveness and ploidy of human mammary carcinomas in short term culture, *Proc. Natl. Acad. Sci. U.S.A.*, 82, 1805, 1985.
170. Smith, H. S., Wolman, S. R., and Hackett, A. J., The biology of breast cancer at the cellular level, *Biochim. Biophys. Acta*, 738, 103, 1984.
171. Laerum, O. D. and Rajewsky, M. F., Neoplastic transformation of fetal rat brain cells in culture after exposure to ethylnitrosourea *in vivo, J. Natl. Cancer Inst.*, 55, 1177, 1975.
172. Roscoe, J. P. and Claisse, P. J., A sequential *in vivo-in vitro* study of carcinogenesis induced in the rat brain by ethylnitrosourea, *Nature (London)*, 262, 314, 1976.
173. Borland, R. and Hand, G. C., Early appearance of "transformed" cells from the kidneys of rats treated with a "single" carcinogenic dose of dimethylnitrosamine (DMN) detected by culture *in vitro, Eur. J. Cancer*, 10, 177, 1974.
174. Park, H. and Koprowska, I., A comparative *in vitro* and *in vivo* study of induced cervical lesions of mice, *Cancer Res.*, 28, 1478, 1968.
175. Koprowska, I., Pertinence of an *in vitro* murine system to early identification of developing human cervical carcinoma, *Acta Cytol.*, 14, 270, 1970.
176. Sanford, K. K., Likely, G. D., Earle, W. R., The development of variations in transplantability and morphology within a clone of mouse fibroblasts transformed to sarcoma-producing cells *in vitro, J. Natl. Cancer Inst.*, 15, 215, 1954.
177. Yamasaki, H., Enomoto, T., Shiba, Y., Kanno, Y., and Kakunaga, T., Intercellular communication capacity as a possible determinant of transformation sensitivity of BALB/c 3T3 clonal cells, *Cancer Res.*, 45, 637, 1985.
178. Lo, K.-Y. and Kakunaga, T., Similarities in the formation and removal of convalent DNA adducts in benzo(a)pyrene-treated BALB/3T3 variant cells with different induced transformation frequencies, *Cancer Res.*, 42, 2644, 1982.
179. Reznikoff, C. A., Bertram, J. S., Brankow, D. W., and Heidelberger, C., Quantitative and qualitative studies of chemical transformation of cloned C3H mouse embryo cells sensitive to postconfluence inhibition of cell division, *Cancer Res.*, 33, 3239, 1973.
180. Heidelberger, C., Freeman, A. E., Pienta, R. J., Sivak, A., Bertran, J. A., Casto, B. C., Dunkel, V. C., Francis, M. S., Kakunaga, T., Little, J. B., and Schechtman, L. M., Cell transformation by chemical agents: a review and analysis of the literature, *Mutation Res.*, 114, 283, 1983.
181. Haber, D. A., Fox, D. A., Dynan, W. S., and Thilly, W. G., Cell density dependence of focus formation in the C3H 10T1/2 transformation assay, *Cancer Res.*, 37, 1644, 1977.

Chapter 13

A MULTISTEP MODEL FOR NEOPLASTIC DEVELOPMENT: ROLE OF GENETIC AND EPIGENETIC CHANGES

J. Carl Barrett

TABLE OF CONTENTS

I.	Introduction	118
II.	Mechanisms of Carcinogenesis In Vivo: Initiation	118
III.	Mechanisms of Carcinogenesis In Vivo: Promotion and Progression	120
IV.	Mechanism of Carcinogenesis In Vitro	121
V.	Immortality	121
VI.	Tumor Suppressor Genes	123
VII.	Conclusion	124
	References	125

I. INTRODUCTION

Any current model of carcinogenesis must consider the multistep nature of neoplastic development. As reviewed in this book by several authors, the evidence for different steps in neoplastic progression comes from diverse studies of pathobiology (Chapter 9), epidemiology (Chapter 10), genetics (Chapter 12), chemical carcinogenesis (Chapters 9 and 11), cell transformation models (Chapter 12), and molecular biology (see Chapter 12 for further discussion. It is now clear that a normal cell must acquire two or more heritable alterations to express tumorigenic potential.[3] Additional changes may be needed for the expression of the malignant phenotype,[5] and neoplastic progression may continue as cells with increasing malignancy and drug-resistant phenotypes arise within the tumor population.[6] Foulds,[7] in his classical discourse on this subject, emphasized the continual nature of tumor progression. Based on the pathological observations of tumors, Foulds developed general principles of tumor progression (Table 1) which are still valid and noteworthy in any discussion on models of neoplastic development.

No model of neoplastic development can be definitive or comprehensive. This is due to the complexity of the carcinogenic process, the multistep nature of the process, the variability among tumors, and the existence of multiple pathways for neoplastic development even for the same tumor type.[7] Nonetheless, with the recent advances in our understanding of the cellular and molecular basis of carcinogenesis, it is possible to speculate on some of the key steps in carcinogenesis. In order to understand the role of genetic and epigenetic changes in carcinogenesis, it is necessary first to define the essential features of cancer cells and to identify the genetic and biochemical factors involved in these changes. A rudimentary understanding of this problem is now emerging and has, in part, been reviewed already in this book. Therefore, in this chapter I will review briefly certain key steps in neoplastic development, discuss the possible genes involved in these steps, and speculate on the role of genetic or epigenetic changes in a multistep model of neoplastic development.

The underlying premise of this model is that cancer is a multistep process and that the cancer cell evolves by clonal expansion of a cell which acquires sequentially different cellular properties. This is certainly not a new hypothesis,[6] and is supported by multiple lines of evidence as discussed in Chapter 12. The major unknown aspect of this model is the cellular and molecular basis for the essential changes in a malignant cell. Insights into this problem can be gained from a discussion of the mechanisms of neoplastic development elucidated from studies of carcinogenesis in vivo and in vitro.

II. MECHANISMS OF CARCINOGENESIS IN VIVO: INITIATION

The process of neoplastic development is often divided into three operationally defined stages — initiation, promotion, and progression.[1,8] This division is useful for discussion purposes, but one should be cautioned not to assume that only three stages exist. Evidence now exists that each of these phases consists of multiple stages.

The process of initiation in the mouse skin carcinogenesis model is an irreversible event induced by a single exposure to a carcinogen.[9,10] These findings led to the hypothesis that a mutational change is involved. Recently Balmain and Pragnell[11] and Balmain and co-workers[12] have reported that the *ras* oncogene is mutated to an activated form in both papillomas and carcinomas of mouse skin. The type of mutations observed depends on the carcinogen used to initiate the tumor, but is independent of the mouse strain or promoter used.[13] This is consistent with the mutation of this oncogene playing a causative role in the initiation process of these tumors. Furthermore,

Table 1
FOULDS' GENERAL PRINCIPLES OF TUMOR PROGRESSION

Rule I	Independent progression of multiple tumors; progression occurs independently in different tumors in the same animal
Rule II	Independent progression of characters; progression occurs independently in different characters in the same tumor
Rule III	Progression is independent of growth; progression occurs in latent tumor cells and in tumors whose growth is arrested; two notable corollaries of Rule III are as follows: At its first clinical manifestation a tumor may be at any stage of progression Progression is independent of the size or clinical duration of a tumor
Rule IV	Progression is continuous or discontinuous by gradual change or by abrupt steps
Rule V	Progression follows one of alternative paths of development
Rule VI	Progression does not always reach an endpoint within the lifetime of the host

From Foulds, L., *Neoplastic Development*, Vol. 1, Academic Press, London, 1969. With permission.

these investigators have recently observed that the Harvey murine sarcoma virus containing the mutated *ras* oncogene can act as an initiator of epidermal tumors of the mouse; the formation of tumors is dependent on subsequent promoter treatment of the animals initiated with the virus.[13] Thus, the mutated form of the *ras* oncogene appears to act in a manner similar to carcinogen treatment which is further support for the hypothesis that this mutation is a key event in the initiation process of mouse skin carcinogenesis.

A role for mutation of the *ras* oncogenes is also indicated in other animal tumor models of chemical carcinogenesis. Zarbl and co-workers[14,15] have reported that the Ha-*ras* oncogene is mutated in a substantial number of mouse mammary carcinomas induced by a single treatment with methylnitrosourea (MNU). This mutagen alkylates DNA at the 0^6 position of guanine which results in mispairing and a transitional mutation of an adenosine to guanosine (G $\rightarrow$ A). This predicted mutation is observed in 48 of 58 of the MNU-induced mammary tumors. However, if the initiator is another carcinogen with a different mutational spectrum, then this specific G $\rightarrow$ A mutation is not reproducibly observed. This implies that the mutation of the *ras* gene is dependent on the initiator and hence is not a spontaneous postinitiation event. Wiseman et al.[16] observed c-Ha-*ras* activation in $B6C3F_1$ mouse hepatocarcinogenesis. The types of mutations observed were consistent with these mutations being an early event resulting directly from the reaction of the ultimate form of the chemical carcinogen with this gene in vivo. However, in other animal model systems the results are not always consistent with *ras* mutation being the initiation event. Guerrero and co-workers[17] have observed mutations of the *ras* oncogene in mouse thymic lymphomas, but the point mutation is the same for chemically and radiation-induced tumors; this is not necessarily the result predicted on the basis of the mutational action of these carcinogens. Radiation-induced point mutations are much less frequent than chemically-induced mutation.[18] This may indicate that in these tumors the *ras* mutations arise at a late stage independent of the initial carcinogen treatments. Alternatively, radiation-induced point mutations may be more relevant to carcinogenesis than predicted.

There are several examples of mutation in the *ras* oncogene occurring late in the carcinogenic process of certain tumors. Vousden and Marshall[19] reported that a metastatic variant of a lymphoma cell line has an activated K-*ras* oncogene while the parent cell line does not, suggesting that the *ras* gene mutation is a late event in the progression of these cells. Gallick et al.[20] observed increased *ras* protein expression in the intermediate stages of progression of colon carcinomas. Studies of human teratocarcinoma cells[21] in culture and chemically-transformed fetal guinea pig cells[22] in culture indicate that the *ras* mutation is a late event in the neoplastic transformation of these cells.

The observations of the *ras* mutation being an early event in some tumors and a late event in others should not be considered contradictory. In fact, the original view of Foulds of neoplastic development as independent progression of characters occurring by alternative pathways (Table 1) predicts such a finding.[7]

The prevailing view at present is that the initiation process involves a mutational event. For the classical two-stage model of mouse skin carcinogenesis, this hypothesis is supported by the increasing evidence that mutation of the *ras* gene is involved in initiation. In other tumor models, mutations of non-*ras* oncogenes may be involved, and future molecular studies should clarify the generality of the mutational basis of the initiation process. However, some authors have recently suggested that existing data are inconsistent with a mutational basis for initiation.[23-25]

Based on measurements of the frequency of initiated cells, a nonmutational mechanism in the induction of certain tumors has been proposed.[23,25] In some models, carcinogens appear to alter a high percentage of the treated cells, increasing their probability to progress to neoplasia. The high frequency of carcinogen-induced initiated cells has been suggested as evidence against a mutational mechanism.[23-25] Nomura[26] has shown that parental exposure of mice to X-rays or urethane induces heritable tumors in a high frequency (>10%) of the offspring. Clifton et al.[24] have estimated on the basis of transplantation experiments that the frequency of initiated cells per clonogenic cells in rat mammary gland exposed to 7-Gy gamma irradiation was 4×10^{-3}.

A high frequency (1 to 100%) for transformation of certain cells in culture has also been observed[3,25] which may have possible relevance to the in vivo observations noted above (although this is by no means clear[3]).

As discussed earlier in Chapter 12, the induction of a high frequency of initiation does not necessarily preclude mutational mechanisms. Several possibilities consistent with mutational events can be offered to explain the high frequency of initial carcinogen-induced cellular changes (e.g., multiple targets, mutational hot spots, and chromosomal mutations[3]). Further studies are needed to define the mechanism(s) of initiation in these models where high frequency events have been indicated. Multiple, carcinogen-induced mechanisms, both genetic and epigenetic, probably exist to increase the probability of a cell acquiring the changes necessary for neoplastic conversion. Different mechanisms should be expected in different systems (with different carcinogens in the same system) and probably even in the same system with the same carcinogen. The elucidation of one mechanism does not preclude alterantive mechanisms in other tumors.

III. MECHANISMS OF CARCINOGENESIS IN VIVO: PROMOTION AND PROGRESSION

The process of tumor promotion undoubtedly involves epigenetic factors. This statement will remain valid regardless of whether or not it can be shown that genetic changes are also involved in tumor promotion. The simplest model of tumor promotion is the clonal expansion of the initiated cells. Farber et al.[1] (Chapter 9) define promotion as "the process whereby an initiated tissue or organ develops focal proliferations (i.e., nodules, papillomas, polyps, etc), one or more of which may act as precursors for subsequent steps in the carcinogenic process." Thus, focal proliferation (i.e., clonal expansion) is clearly an essential part of promotion. The mechanisms by which clonal expansion can occur are varied, and hence the mechanisms of tumor promotion must be multiple.[1,27] Differential growth of the initiated cell vis-à-vis the normal cells in a tissue can result from a differential growth stimulus of the initiated cell, decreased terminal differentiation or cell death of the initiated cell, or resistance

REFERENCES

1. Farber, E., Rotstein, J. B., and Ericksson, L. C., Cancer development as a multistep process: experimental studies in animals, in *Mechanisms of Environmental Carcinogenesis,* Vol. 2, Barrett, J. C., Ed., CRC Press, Boca Raton, Fla., 1987, chap. 9.
2. Kaldor, J. M. and Day, N. E., Interpretation of epidemiological studies in the context of the multistage model of carcinogenesis, in *Mechanisms of Environmental Carcinogenesis,* Vol. 2, Barrett, J. C., Ed., CRC Press, Boca Raton, Fla., 1987, chap. 10.
3. Barrett, J. C. and Fletcher, W. F., Cellular and molecular mechanisms of multistep carcinogenesis in cell culture models, in *Mechanisms, of Environmental Carcinogenesis,* Vol. 2, Barrett, J. C., Ed., CRC Press, Boca Raton, Fla., 1987, chap 12.
4. Hennings, H., Tumor promotion and progression in mouse skin, in *Mechanisms of Environmental Carcinogenesis,* Vol. 2, Barrett, J. C., Ed., CRC Press, Boca Raton, Fla., 1987, chap. 11.
5. Fidler, I. J. and Hart, I. R., The development of biological diverstiy and metastatic potential in malignant neoplasma, *Oncodev. Biol. Med.,* 4, 161, 1982.
6. Nowell, P. C., The clonal evolution of tumor cell populations, *Science,* 194, 23, 1976.
7. Foulds, L., *Neoplastic Development,* Vol. 1, Academic Press, London, 1969.
8. Pitot, H. C., Goldsworthy, T., and Moran, S., The natural history of carcinogenesis: implication of experimental carcinogenesis in the genesis of human cancer, *J. Supramol. Struct. Cell. Biochem.,* 17, 133, 1981.
9. Boutwell, R. K., The function and mechanism of promoters of carcinogenesis, *CRC Crit. Rev. Toxicol.,* 2, 419, 1974.
10. Loehrke, H., Schweizer, J., Dederer, E., Hesse, B., Rosenkranz, G., and Goerttler, K., On the persistence of tumor initiation in two-stage carcinogenesis on mouse skin, *Carcinogenesis,* 4, 771, 1983.
11. Balmain, A. and Pragnell, I. B., Mouse skin carcinomas induced *in vivo* by chemical carcinogens have a transforming Harvey-*ras* oncogene, *Nature (London),* 303, 72, 1983.
12. Balmain, A., Ramsden, M., Bouden, G. T., and Smith, J., Activation of the mouse cellular Harvey-*ras* gene in chemically induced benign skin papillomas, *Nature (London),* 307, 658, 1984.
13. Balmain, A., personal communication.
14. Zarbl, H., Sukumar, S., Martin-Zanca, D., Santos, E., and Barbacid, M., Molecular assays for detection of *ras* oncogenes on human and animal tumors, in *Carcinogenesis — A Comprehensive Survey,* Vol. 9, Barrett, J. C. and Tennant, R. W., Eds., Raven Press, New York, 1985, 1.
15. Zarbl, H., Sukumar, S., Arthur, A. V., Martin-Zanca, D., and Barbacid, M., Direct mutagenesis of Ha-*ras*-1 oncogenes by N-nitroso-N-methylurea during initiation of mammary carcinogenesis in rats, *Nature,* 315, 382, 1985.
16. Wiseman, R. W., Stowers, S. J., Miller, E. C., Anderson, M. W., and Miller, J. A., Activating mutations of the c-Ha-*ras* proto-oncogene in chemically induced hepatomas of the male B6C3F$_1$ mouse, *Proc. Natl. Acad. Sci. U.S.A.,* 83, 5825, 1986.
17. Guerrero, I., Vilasante, A., Corces, V., and Pellicer, A., Activation of a c-K-*ras* oncogene by somatic mutation in mouse lymphomas induced by gamma radiation, *Science,* 225, 1159, 1984.
18. Thacker, J. and Cox, R., The relationship between specific chromosome aberrations and radiation-induced mutations in cultured cells, in *Radiation-Induced Chromosome Damage in Man,* Alan R. Liss, New York, 1983, 235.
19. Vousden, K. H. and Marshall, C. J., Three different activated *ras* genes in mouse tumours; evidence for oncogene activation during progression of a mouse lymphoma, *EMBO J.,* 3, 913, 1984.
20. Gallick, G. E., Kurzrock, R., Kloetzer, W. S., Arlinghaus, R. B., and Gutterman, J. U., Expression of p21^{ras} in fresh primary and metastic human colorectal tumors, *Proc. Natl. Acad. Sci. U.S.A.,* 82, 1795, 1985.
21. Tainsky, M. A., Cooper, C. S., Giovanella, B. C., and Vande Woude, G. F., An activated *ras^N* gene: detected in late but not early passage human PAI teratocarcinoma cells, *Science,* 225, 643, 1984.
22. Doniger, J., Molecular characterization of oncogenes in guinea pig lines chemically initiated *in vitro*: acquisition of tumorigenecity is associated with activated *ras* related oncogenes, in *Carcinogenesis — A Comprehensive Survey,* Vol. 9, Barrett, J. C. and Tennant, R. W., Eds., Raven Press, New York, 1985, 51.
23. Gould, M. N., Radiation initiation of carcinogenesis *in vivo*: a rare or common cellular event, in *Radiation Carcinogenesis: Epidemiology and Biological Significance,* Boice, J. D., Jr. and Frawmeniz, J. F., Jr., Eds., Raven Press, New York, 1984, 347.
24. Clifton, K. H., Tanner, M. A., and Gould, M. N., Assessment of radiogenic cancer initiation frequency per clonogenic rat mammary cell *in vivo, Cancer Res.,* 1986, in press.

25. Kennedy, A. R., Evidence that the first step leading to carcinogen-induced malignant transformation is a high frequency, common event, in *Carcinogenesis — A Comprehensive Survey*, Vol. 9, Barrett, J. C. and Tennant, R. W., Eds., Raven Press, New York, 1985, 355.
26. Nomura, T., Parental exposure to x-rays and chemicals induces heritable tumours and anomalies in mice, *Nature (London)*, 296, 575, 1982.
27. Barrett, J. C. and Thomassen, D. G., Use of quantitative cell transformation assays in risk estimation, in *Methods of Estimating Risk in Human and Chemical Damage in Non-human Biota and Ecosystems, SCOPE, SGOMSEC 2, IPCS Joint Symposia 3*, Vouk, V. B., Butler, G. C., Hoel, D. G., and Peakall, D. B., Eds., John Wiley & Sons, New York, 1985, 201.
28. Aldaz, C. M., Conti, C. J., Fries, J. W., Klein-Szanto, A. J. P., and Slaga, T. J., Chromosomal abnormalities in mouse skin tumors induced by two stage carcinogenesis, *Proc. Am. Assoc. Cancer Res.*, 26, 136, 1985.
29. Smith, J. R. and Hayflick, L., Variation in the life-span of clones derived from human diploid cell strains, *J. Cell Biol.*, 62, 48, 1974.
30. Smith, J. R. and Whitney, R. G., Intraclonal variation in proliferative potential of human diploid fibroblasts: stochastic mechanism for cellular aging, *Science*, 207, 82, 1980.
31. Smith, J. R., Pereira-Smith, O., and Good, P. I., Colony size distribution as a measure of age in cultured cells: a brief note, *Mech. Aging Dev.*, 6, 283, 1977.
32. Martin, G. M., Cellular aging — clonal senescence, *Am. J. Pathol.*, 89, 489, 1977.
33. Martinez, A. O., Norwood, T. H., Prothero, J. W., and Martin, G. M., Evidence for clonal attenuation of growth potential in Hela cells, *In Vitro*, 14, 996, 1978.
34. Sheldrake, A. R., The aging, growth and death of cells, *Nature (London)*, 250, 381, 1974.
35. Buick, R. N. and Pollak, M. N., Perspectives on colongenic tumor cells, stem cells, and oncogenes, *Cancer Res.*, 44, 4909, 1984.
36. Weissman, B. E., Suppression of tumorigenicity in mammalian cell hybrids, in *Mechanisms of Environmental Carcinogenesis*, Vol. 1, Barrett, J. C., Ed., CRC Press, Boca Raton, Fla., 1987, chap. 3.
37. Mulvihill, J. J., Genetic repertory of human neoplasia, in *Genetics of Human Cancer*, Mulvihill, J. J., Miller, R. W., and Fraumeni, J. F., Eds., Raven Press, New York, 1971, 137.
38. Knudson, A. G., Jr., Hereditary cancer, oncogenes, and antioncogenes, *Cancer Res.*, 45, 1437, 1985.
39. Cavenee, W. K., Dryja, T. P., Phillips, R. A., Benedict, W. F., Godbout, R., Gallie, B. L., Murphree, A. L., Strong, L. C., and White, R. L., Expression of recessive alleles by chromosomal mechanisms in retinoblastoma, *Nature (London)*, 305, 779, 1983.
40. Murphree, A. L. and Benedict, W. F., Retinoblastoma: clues to human oncogenesis, *Science*, 223, 1028, 1984.
41. Hethcote, H. W. and Knudson, A. G., Model for the incidence of embryonal cancers: application to retinoblastoma, *Proc. Natl. Acad. Sci. U.S.A.*, 75, 2453, 1978.
42. Knudson, A. G. and Strong, L. C., Mutation and cancer: a model for Wilms' tumor of the kidney, *J. Natl. Cancer Inst.*, 48, 313, 1972.
43. Knudson, A. G. and Meadows, A. T., Developmental genetics of neuroblastoma, *J. Natl. Cancer Inst.*, 57, 675, 1976.
44. Knudson, A. G. and Meadows, A. T., Developmental genetics of neural tumors in man, in *Cell Differentiation and Neoplasia*, Proc. 30th Ann. Symp. on Fundamental Cancer Res., Saunders, G. F., Ed., Raven Press, New York, 1978, 83.
45. Hansen, M. F., Koufos, A., Gallie, B. L., Phillips, R. A., Fodstad, O., Brogger, A., Gedde-Dahl, T., and Cavenee, W. K., Osteosarcoma and retinoblastoma: a shared chromosomal mechanism revealing recessive predisposition, *Proc. Natl. Acad. Sci. U.S.A.*, 82, 1, 1985.
46. Koufos, A., Hansen, M. F., Lampkin, D. B., Workman, M. L., Copeland, N. G., Jenkins, N. A., and Cavenee, W. K., Loss of alleles at loci on human chromosome 11 during genesis of Wilms' tumour, *Nature (London)*, 309, 170, 1984.
47. Fearon, E. R., Feinberg, A. P., Hamilton, S. H., and Vogelstein, B., Loss of genes on the short arm of chromosome 11 in bladder cancer, *Nature (London)*, 318, 377, 1985.
48. Rogler, C. E., Sherman, M., Su, C. Y., and Shafritz, D. A., Deletion in chromosome 11p associated with a hepatitis B integration site in hepatocellular carcinoma, *Science*, 230, 319, 1985.
49. Eker, R. and Mossige, J., A dominant gene for renal adenomas in the rat, *Nature (London)*, 189, 858, 1961.
50. Eker, R., Mossige, J., Johannessen, J. V., and Aars, H., Hereditary renal adenomas and adenocarcinomas in rats, *Diagn. Histopathol.*, 4, 99, 1981.
51. Grateff, E., Malignant neoplasms of genetic origin in *Drosophila melanogaster*, *Science*, 200, 1448, 1978.
52. Anders, F., Schartl, M., Barnekow, A., and Anders, A., *Xiphophorus* as an *in vivo* model for studies on normal and defective control of oncogenes, *Adv. Cancer Res.*, 42, 191, 1984.

INDEX

A

2-AAF, see 2-Acetylaminofluorene
abl gene, 78
Acetone, 66—69
2-Acetylaminofluorene (2-AAF), 3
Adenocarcinoma, 81
Adenovirus, 99, 103
Adenovirus E1a gene, 78, 87, 97
Age-incidence curve, 24—25, 74
    estimation of, 26—27
    multistage carcinogenesis and, 26—29
Alcohol consumption, 33
o-Aminoazotoluene, 3
4-Aminobiphenyl, 46
Anchorage-independent growth, 76, 85, 87, 102
    in human cells, 103—104
    mechanisms of, 93—98
    in rodent cells, 103—104
Aneuploidy, 52, 88, 94, 105, 121
Angiosarcoma, 47
Ankylosing spondylitis, 38—40
Antioncogenes, 50
Apoptosis, 7, 121, 124
Arsenic, 46, 80
Asbestos, 33, 40—44, 52, 80, 104
Atomic bombing, 30, 37—40
ATPase, 4
Azo dyes, 3

B

BALB/c 3T3 cells, 92—93, 99
BaP, see Benzo(a)pyrene
Benzene, 47
Benzidine, 46
Benzo(a)pyrene (BaP), 37, 74—75, 92, 104
Benzoyl peroxide, 70
BHK cells, 94
Birth cohort, 26—27
Bis-chloromethyl ether, 47
Bladder cancer, 37, 123
Bloom's syndrome, 102—103
B lymphocytes, 88
BP6T cells, 98
Brain, 91
Breast cancer, 24—30, 39, 47—49
Breast tissue aging, 26
5-Bromodeoxyuridine, 79
Bronchial carcinoma, 42
Bronchial epithelial cells, human, 85—86, 100, 104
Burkitt's lymphoma, 29, 31, 51

C

C3H 10T1/2 cells, 92—93, 96, 99
CAK cells, 93—94
Cancer Incidence in Five Continents, 26
Cancer registries, 26, 28
Carcinogen
    activation of, 5
    exposure to
        dose of, 24—35
        duration of, 24—35
        multiple carcinogens, 34—35
Carcinogenesis
    human, see also Epidemiological studies, 22
    in vitro, 121
    multistage, see Multistage carcinogenesis
    overall pattern of, 14
Cell-cell interactions, 65, 93, 124
Cell culture, see also specific types of cells
    early stages in transformation of, 79—96
        epithelial cells, 81—86
        immortality, 86—91
        Syrian hamster embryo cells, 79—81
    late stages in transformation of, 91—101
    multistage carcinogenesis in, 74—77, 121
Cell cycle, 9
    altered shut-off of, 10
    quantal, 13
Cell death, 7—10, 13, 15, 90, 121, 124
Cell differentiation, 13, 83—84, 124
    defects and neoplastic transformation, 99
    terminal, 65, 82—85, 90
Cell differentiation genes, 124
Cell division, density-dependent inhibition of, 80—81
Cell evolution, 13
Cell hybrids, 90, 94, 98—99, 104
Cell immortalization, 121—123
    in cell culture carcinogenesis, 74—80, 84—91, 100—101
    in human cells, 103—105
    in rodent cells, 103—105
    spontaneous, 86—87
Cell proliferation
    in cell culture carcinogenesis, 88, 90, 100
    defects in, 99
    in hepatocellular carcinogenesis, 5—10, 12—15
    inhibition of, 12
    in mouse skin carcinogenesis, 62, 65
Cell senescence, 74, 86, 90, 121
Cell transformation, see also Morphological transformation; Neoplastic transformation, 75, 78
    tumor promoters in, 99—100
Cervical cancer, 38—39, 51
Cervix, 91
c-Ha-*ras*-gene, 119
CHEF cells, 102
    16-2, 94—95, 97
    18-1, 94—97
Childhood tumors, 24—25
Chinese hamster cells, 100

Chinese hamster embryo fibroblasts, see CHEF cells
bis-Chloromethyl ether, 47
Choline, 3—4
Chondrocytes, 78
Chromate products, 47
Chromium, 47
Chromosome aberrations, 51—52, 124
    in cell culture carcinogenesis, 79—80, 95—96, 98, 105
Cigarette smoking, 28—30, 33, 36—37, 42—43, 46, 52
Clonal evolution, 13
Clonal expansion, 120—121, 124
    in cell culture carcinogenesis, 74
    in hepatocellular carcinogenesis, 13—14
    in mouse skin carcinogenesis, 61—63
Clonogenicity, 122
c-mos gene, 105
c-myc gene, 51, 87
Colon cancer, 50—51, 74, 119
Confluence, 92—93
Confounding, 35—36
Connective tissue cells, C3H mouse, 92
Contact inhibition, 80—81, 85
Contact-insensitive cells, 80—81
Copper smelter workers, 46
Crisis, 86, 90, 96
Cycasin, 5
Cyproterone acetate, 3
Cytochrome P-450, 8

D

DDT, 3—4
Dibromoethane, 5
Dichloroethane, 5
Diethylstilbestrol, 80
p-Dimethylaminoazobenzene, 3
7,12-Dimethylbenz(a)anthracene (DMBA), 60—64, 66—69
DMBA, see 7,12-Dimethylbenz(a)anthracene
DNA
    changes and morphological transformation, 79
    lesions and tumor initiation, 11—12
    methylation of, 124
    transfection of, 77, 98, 101
DNA adducts, 5—6
DNase I, 79
Drosophila melanogaster, 123—124
DT-diaphorase, 4—5, 8
Dyestuff industry, 46—47
Dysplastic nevus syndrome, 50

E

EJ-ras gene, 100—103
Embryo cells
    mouse, 101
    rat, 93, 99

Syrian hamster, 74—81, 88—89, 97—101, 103
EMS, see Ethyl methane sulfonate
Endometrial cancer, 47—50
Epidemiological studies, 22—23
    of arsenic, 46
    of asbestos, 40—44
    of cigarette smoking, 36—37
    of hormones as carcinogens, 47—49
    of nickel, 44—46
    of radiation exposure, 37—40
    statistical issues in interpretation of, 35—36
Epidermal carcinogenesis, 60
Epidermal cells, mouse, 82—84
Epidermal growth factor, 87
Epigenetic changes, 117—124
Epithelial cancer, 22
Epithelial cells, 77, 86, 104
    bronchial, 85—86, 100, 104
    human, 89, 104
    mammary, 89, 104
    multistage carcinogenesis in, 81—86
    rat tracheal, see Rat tracheal epithelial cells
Epithelial foci, 81
Epoxide hydrolase, 4, 8
Epstein-Barr virus, 51, 87—88
erbB gene, 78
Esophageal cancer, 37
Estrogen replacement therapy, 48—50
Estrogens, 47—49, 52
Ethyl methane sulfonate (EMS), 95
N-Ethyl-N-nitrosourea, 80

F

Familial clustering of cancers, 50
Familial polyposis, 50
fes gene, 78
Fibrinolytic activity, 75—76
Fibroblast growth factor, 104
Fibroblasts, 90
    chicken embryo, 101
    Chinese hamster embryo, see CHEF cells
    fetal guinea pig, 96
    human, 86, 102—104, 121
    mouse, 77
    rat, 77—78, 88
    rodent, 101—102
Fibrosarcoma, 74—75
Fish, hybrid, 123
Fluocinolone acetonide, 68
fms gene, 78
Foci of altered hepatocytes, see Hepatocyte foci
FOL-2-cells, 94
fos gene, 78
fps gene, 78

G

G_1 state, 99

Gamma irradiation, 102, 120
G_D state, 99
Gene activation, 11—12
Gene amplification, 124
Gene dosage, 99
Gene expression, 61—63, 65, 78
Gene rearrangement, 11—12
Genetic changes, 117—124
Glucose-6-phosphatase, 4
β-Glucosidase, 5
γ-Glutamyltransferase, 4, 8
Glutathione, 5, 8
Glutathione-S-transferase, 8
Growth control processes, 88
Growth factors
    in cell culture carcinogenesis, 78, 84, 87—88
    in hepatocellular carcinogenesis, 13, 15
    in mouse skin carcinogenesis, 65
Growth regulatory genes, 122—123

H

Hamster cells, 87, 94
Ha-*ras* gene, 80, 84—85, 96, 100—101, 119
Harvey murine sarcoma virus, 77, 84, 101, 119
Hazard function, 24
Heart mesenchymal cells, 87
HeLa cells, 104
Hepatitis B virus, 2, 29
Hepatoblastoma, 123
Hepatocellular carcinogenesis, 119
    comparison with other organ carcinogenesis, 14
    models of, 2—3
    tumor initiation in, 4—6
    tumor progression in, 8—10
    tumor promotion in, 6—8
Hepatocellular carcinoma, 29, 123
Hepatocyte, initiated, 4
Hepatocyte foci, 3, 6—7
Hepatocyte nodule, 3—4, 7—8
    conversion to carcinoma, 10—11
    persistence of, 7—9
    remodeling of, 7
    transplantation to spleen, 10
α-Hexachlorocyclohexane, 3
Hodgkin's disease, 24
Hormones, carcinogenicity of, 47—49
HT 1080 cells, 98—99
Human cells, 86—87, 89—90
    transformation in, 101—105
    tumor, 98—99
    tumorigenicity in, 102—103
Human papilloma virus, 51
Human tumors, 91
Hyperplasia, 63, 124

I

Immortalizing genes, 78, 100—101
Immune surveillance, 52
Incidence function, 24
Initiation, see Tumor initiation
Interleukin, 87
Invasive properties, 11, 13—14
Ionizing radiation, see Radiation exposure

J

JB-6 cells, 99

K

Keratin, 83
Keratinocytes, 84—85, 90, 103
Kidney, 91
Ki-*ras* gene, 103
Kirsten sarcoma virus, 84
K-*ras* gene, 80, 119

L

Laryngeal cancer, 28, 37
Lesch-Nyhan cells, 104
Leukemia, 26, 37—40, 47
    chronic myelocytic, 51
Liver cell cancer, see Hepatocellular carcinogenesis
Liver cells, rat, 89
Long terminal repeat, 101
Lung cancer, 26—29, 35—37, 40—47
    metastatic from skin, 69
Lymph node metastasis, 69
Lymphoma
    Burkitt's, 29, 31, 51
    thymic, 119
Lymphoma cells, 119

M

Malignant conversion, 60—61, 65—68, 70
Mammary epithelial cells, 89, 104
Mammary tumor, 119
Melanoma, 14, 26—28, 50—51, 103
Menarche, 48
Menopause, 47—49
Mercapturic acid, 8
Mesothelial cells, 104
Mesothelioma, 40—44, 52
Metastasis, 11, 13—14, 61, 69—70
Methionine, 3—4
3-Methylcholanthrene, 74, 93, 96
N-Methyl-*N'*-nitro-*N*-nitrosoguanidine (MNNG), 65—69, 81, 94—96, 104
Methylnitrosourea (MNU), 119
Mezerein, 60—64, 69—70
mil gene, 78
Mineral fiber, 80

Mito-inhibition, 4
Mixed function oxidase system, 5, 8
MNNG, see N-Methyl-N'-nitro-N-nitrosoguanidine
MNU, see Methylnitrosourea
Monosomy, 97—98
Morphological transformation, 75—81, 88—89, 94
   in human cells, 103
   one-hit model for, 79
   reversion of, 93
   in rodent cells, 103
   suppression of, 93
mos gene, 78
Mouse
   BALB/c, 83
   Charles River CD-1, 63, 65, 67—70
   NMRI, 62
   SENCAR, 60—68, 70, 83
Mouse cells, 87, 92—93
Mouse mammary tumor virus, 87
Mouse skin carcinogenesis, 22, 59—70, 118—121
   malignant conversion, 65—68
   mechanisms of selection, 65
   metastasis, 69
   two-stage promotion, 60—64
Multihits, 15
Multistage carcinogenesis, 2, 22—23
   age-incidence curves and, 26—29
   carcinogen exposure and, 29—49
   in cell culture, see also Cell culture; specific types of cells, 73—106
   epigenetic changes in, 117—124
   in epithelial cells, 81—86
   genetic changes in, 117—124
   mathematical formulation of, 23—26
   mouse skin, see also Mouse skin carcinogenesis, 22, 59—70, 118—121
   oncogenes and, 77—79
   precursor states and, 49—51
Mutagen, 79
Mutation, 50, 118—120, 123
   in cell culture carcinogenesis, 79, 88
   in hepatocellular carcinogenesis, 11
   point, 80, 124
Mutational hot spot, 79
myb gene, 78, 87
myc gene, 77—78, 87—89, 97—101

N

β-Naphthylamine, 46
Nasal sinus cancer, 44—46
Nasopharyngeal cancer, 29, 31
Neoplastic development, 117—124
Neoplastic transformation, 74
   in cell culture
      early stages of, see also Cell culture, 79—96
      comparison of human and rodent cells, 101—105
      differentiation defects and, 99
      late stages in, 91—101
      single-step, 100—101
      spontaneous, 92
neu gene, 78
Neuroblastoma, 123
Nevi, 50
Nickel, 44—46, 52
NIH/3T3 cells, 77, 96—97, 102
Nitrofuran, 5
4-Nitro-quinoline-*N*-oxide (4-NQO), 5, 65—70, 76, 95, 102—103
N-Nitrosomorpholine, 3
N-*myc* gene, 78, 87
Nodule, 5—6, 14, 120
   hepatocyte, see Hepatocyte nodules
4-NQO, see 4-Nitro-quinoline-*N*-oxide
N-*ras* gene, 80
Nucleic acid analogs, 79

O

Oncogenes, see also specific genes, 13, 50—51, 121—124
   in cell culture carcinogenesis, 74, 90, 96—97
   in cell immortalization, 87—88
   in morphological transformation, 80
   multistage carcinogenesis and, 77—79
Oophorectomy, 48—49
Oral cavity cancer, 37
Oral contraceptives, 48
Ornithine decarboxylase, 65
Orotic acid, 4
Osteosarcoma, 123
Ovarian cancer, 24, 28

P

p53 gene, 78, 87—88, 97, 100—101
Pancreatic cancer, 37
Papilloma, 5—6, 14, 22, 83—84, 118—121
Papilloma phenotype, expression of, 61—63
PCB, 3—4
PDGF, see Platelet-derived growth factor
Phenobarbital, 3—4
Pheochromocytoma, 123
Phorbol ester, 121
Platelet-derived growth factor (PDGF), 87
Polycyclic hydrocarbons, 5, 79, 96, 124
Polyoma large T gene, 78, 87—88
Polyoma middle T gene, 78
Polyp, 5—6, 14, 50, 120
Precursor states, 49—51
Preneoplastic cells, 49—51
   in cell culture carcinogenesis, 75—76, 81—82, 95
   in hepatocellular carcinogenesis, 6—7
Preneoplastic lesions, 121
Progression, see Tumor progression
Promotion, see Tumor promotion
Propane sultone, 104
Prostaglandin synthesis, 5, 65

Prostatic cancer, 28
Protooncogenes, see also specific genes, 77

R

Radiation exposure, 30, 33, 37—40, 52, 119
Radiation therapy, 38—39, 48—49
Radiologists, 37
raf gene, 78
ras gene, 77—80, 84—86, 96—101, 104, 118—120
Rat cells, 87
Rat tracheal epithelial (RTE) cells, 85, 89—90, 96
  enhanced-growth variants of, 81—82, 89
Rauscher leukemia virus, 93
Rectal cancer, 74
REF-52 cells, 97
Regulatory genes, 12
Renal cancer, 123
Resistant hepatocyte model, 3—4, 12
Retinoblastoma, 50, 74, 99, 123
Retinoic acid, 68
Retrovirus, 77
Rhabdomyosarcoma, 123
Rodent cells, transformation in, 101—105
Rous sarcoma virus (RSV), 101
RSV, see Rous sarcoma virus
RTE cells, see Rat tracheal epithelial cells

S

Simian virus 40, 89—90, 103—104
Simian virus 40 large T gene, 78, 87
ski gene, 78
Skin carcinogenesis, see Mouse skin carcinogenesis
Smelter workers, 46
Solid tumors, radiation-induced, 37—40
S phase, 9
Squamous cell carcinoma, see also Mouse skin carcinogenesis, 59—70, 81
  metastasis of, 69—70
src gene, 78, 97, 101
Stem cells, 25, 81, 89—91, 122—124
Stop model of hepatic carcinogenesis, 3
Synthetic dye industry, 46
Syrian hamster cells, 87—88, 94
Syrian hamster embryo cells, 74—81, 88—89, 97—101, 103

T

3T3-T cells, 99
Teleocidin, 70
Teratocarcinoma cells, 119
Testicular cancer, 24
12-*O*-Tetradecanoylphorbol-13-acetate (TPA), 60—61, 63—70, 83—84, 99—100
Thyroid cancer, 26

Toxicity of carcinogen, 79
TPA, see 12-*O*-Tetradecanoylphorbol-13-acetate
Trachea, 91
Tracheal epithelial cells, rat, see Rat tracheal epithelial cells
Transcriptional enhancer, 100
Transfection, 77, 98, 101
Transformation, see Cell transformation; Morphological transformation; Neoplastic transformation
Transforming oncogenes, 78
Trisomy, 121
Tumorigenicity, 75—78, 85, 87, 94—101
  in human cells, 102—103
Tumor initiation, 2—3, 14, 22, 82—84
  biochemical lesions in, 5
  fixation by cell proliferation, 5—6
  in hepatocellular carcinogenesis, 4—6
  mechanisms of, 11—12, 118—120
  in mouse skin carcinogenesis, 60
Tumor progression, 2, 15, 118—119
  in hepatocellular carcinogenesis, 8—10
  mechanisms of, 13—14, 120—121
  in mouse skin carcinogenesis, 59—70
Tumor promoter
  in cell transformation, 99—100
  incomplete, 63
  selection for initiated cells by, 65
Tumor promotion, 2—3, 14—15, 22
  delayed, 61
  foci of altered cells in, 6—7
  in hepatic carcinogenesis, 6—8
  mechanisms of, 12—13, 120—121
  in mouse skin carcinogenesis, 59—70
  nodule formation in, 7—8
  reversibility of, 60, 63
  two-stage, 60—64
Tumor suppression genes, 98—99, 105, 121—124

U

UDP-glucuronyltransferase I, 8
Underground miners, 37
Urethane, 65—70, 120
Uterine cancer, 28
UV irradiation, 79, 92, 99

V

v-Ha-*ras* gene, 77, 97, 101, 103
Vinyl chloride, 47
v-*myc* gene, 87
v-*src* gene, 101

W

Weibull distribution, 24
Wilms' tumor, 50, 99, 123

X

Xiphophorus hybrid, 123—124
X-irradiation, 120

Y

yes gene, 78